# A FACE
# FOR ME

# A FACE FOR ME

## Debbie Diane Fox

with Jean Libman Block

*Wyden Books*

*Manufactured in the United States of America*

*Second Printing*

*Trade distribution by Simon and Schuster*
*A Division of Gulf + Western Corporation*
*New York, New York 10020*

*Designed by Tere LoPrete*

---

**Library of Congress Cataloging in Publication Data**

Fox, Debbie Diane, 1955–
    A face for me.

      1. Face—Abnormalities—Biography. 2. Fox, Debbie, Diane, 1955–    I. Block, Jean Libman, joint author. II. Title.
QM695.F32F69    362.7′8′19752 [B]    78–10266
ISBN 0-88326-156-1

---

# Contents

# A FACE
# FOR ME

*Now faith is the substance of things hoped for, the evidence of things not seen.*

HEBREWS 11:1

# ∽ 1 ∾

# *The Mirror Told Me*

I never saw my face until I was eight years old. I guess I knew I was different from other children. I'd always known that. It was because I was different that I couldn't go out and play with the boys and girls who went to fish and splash in Soddy Creek, which ran past our house. When I tried to go play with the children who were laughing and having fun, Mama put a gentle hand on my shoulder and held me back. Mama said, "Debbie, that's not for you. You stay right here and play up on the porch."

Sometimes I slipped away from her and ran partway down to the brook. The children would stop what they were doing and stare up at me. Then they'd whisper together. Or they'd make faces at me—ugly, scary faces. I wanted so much to be with those children down there! But something always stopped me. Not just Mama's hand on me, but something inside me. I always went back to the house—not really wanting to go inside, but knowing I had to.

I thought it was my hand that kept me away from the children. My right hand wasn't a hand at all, just a stump

—no palm, no fingers. I'd always known my hand was no good, and I hated it. I'd asked Mama a dozen times what was wrong with it. She told me the same thing every time: "It's for a reason, Debbie. God made your hand that way for a reason. Someday maybe it'll be better. But if it isn't, you'll have to accept it. God made it that way for a reason."

I really thought it was my hand that made me different. I didn't know about my face yet. Then one day I found a hand mirror in the bathroom, tucked in back of the toilet tank. I idly picked it up.

I'm not sure I even knew what a mirror was at that time. As a child, you take what you're given; you don't go out searching for what you've never heard about. Our house has always been neat and plain. Back then, there weren't many extras in it. Since I've grown up, I've been invited into some fancy homes and I've seen whole doors covered with mirrors and bathrooms with mirrors all the way around. We never lived like that. Even at my sister Joyce's and at my brothers' homes, there was not so much vanity that many mirrors were needed. Our Church of God religion doesn't let us wear make-up and jewelry, so there isn't cause for primping in mirrors. So don't be surprised when I tell you I never thought to climb up to look in the mirror in the bathroom, which Daddy used for shaving, or the one over the high chest in Mama and Daddy's bedroom.

Now I took this hand mirror and looked in it. When I saw my face, I could feel a scream come tearing out of my throat. My face was a terrible sight. It wasn't like any face I'd ever seen. My nose wasn't a nose at all—it was just a shapeless flap of flesh that seemed to hang on my face at eye level. Instead of nostrils, there were two gaping holes. My eyes were way over on the side of my head, so far apart that the center of my face looked empty and blank. The left eye was misshapen and had only part of an upper lid. The eyebrows were funny-looking tufts of hair that went up, down, and across. My upper lip had jagged scars on it. It was very small and tight—almost too small to cover the two stubby front teeth that stuck out, straight forward, nearly at a right angle to the way they should be.

My eyes were so far apart that I had to move the mirror from one side to the other so I could take turns looking in it with each eye. I couldn't see myself by looking straight ahead in the glass.

I stifled the screaming. But then the tears came—hot, terrible, scalding tears. So that was what I looked like! That was why I couldn't play with the other children, go to school, go to church, run into the store to buy candy or ice cream. All these things had been forbidden to me. Because of my hand, I'd always thought. Now I knew. I was ugly. I was horrible. I wasn't fit to look at.

I put the mirror down and went into the other room to confront Mama. I guess she knew what was coming from the way I looked at her, and from my red, wet eyes. "Mama," I cried, "how could you? Why didn't you tell me? Why didn't I know?"

Mama is a very beautiful woman, very dignified, with heavy hair that I've always remembered as silvery gray, twisted into a kind of crown on her head. She took me into her arms and held me tight. I could feel her hand stroking my hair. "We didn't want to hurt you, Debbie. We've never wanted others to hurt you."

"But you should have told me," I sobbed.

"Maybe we should have." She sighed. "But what good would it do? God made you this way, and we have to believe it's for a reason, His reason."

By the time Daddy came home that night, I'd calmed down. I didn't want Daddy to see me upset. Daddy was special and I loved him in a special way. It was all right for Mama to see me crying and upset, but I had to be brave for Daddy. Of course, Daddy knew right away what had happened. He took me on his lap and rubbed his forehead against mine, and I felt soothed and loved. "We're going to get your face fixed for you, baby," he told me softly. "It's going to be fixed. Just believe and pray. It's going to be fixed."

Once I knew what was wrong, I began to put the pieces together. There were all the visits to Dr. Barnwell, the tall, gray-haired doctor in Chattanooga. Now I realized he was a plastic surgeon and had already started patching up my face. It had been worse—much worse, if you can believe it—when I was born. And Dr. Phillips, the friendly dentist. He was working on my teeth and jaw. They really were trying to fix me up. They were trying to make me look better.

Still, it hurt that everyone had known but me. Not just Mama and Daddy, but my sister, Joyce, and my two big brothers, Ralph and Bobby, and Mrs. Apple, who came to the house to teach me, and the children down at the creek I couldn't play with, and Brother Hickman, the Church of God preacher who lived across the way. They all knew. They knew about my awful, deformed face and I didn't. Even strangers who saw me knew. That rankled—it hurt deep down.

But then, as I slowly settled into knowing what was really wrong with me and struggled to accept my difference, I began to be more forgiving of my parents. Maybe they'd been right after all to keep my deformity a secret from me. Looking back now, I realize the discovery really meant an end to my innocence. Before, even though I couldn't play with the other children, I could imagine myself among them. In my fantasies I was always surrounded by boys and girls my own age. We were jumping rope, splashing in the brook, walking to the five-and-ten, whispering, giggling, linking arms.

Now these fantasies were dead. How could I expect to be accepted with a face like mine? It was crazy to think anyone would ever want to look at it—I could hardly look at it myself! So I was alone. A misfit. A shut-away. A freak. All alone except for my family. They loved me enough not to care how I looked. Gradually I began to see Mama and Daddy and my sister and brothers through new eyes.

I began to think about how terrible it must have been for them—first to accept that I was really their flesh and blood, then to have to protect me and defend me when other people were nosy or surprised or plain cruel. How much they must have suffered for me! No wonder they hadn't wanted me to know. Every day I didn't know about myself was another day of peace for me. Once I knew, there was no more peace. Each new person I had to meet was a new ordeal. Each sidelong glance at me or look of horror was another hurtful slap.

Only God could look at me without shuddering, for He had made me. He had made me this way for a reason. I had to hold fast to that belief. It was my only hope. There *had* to be a reason. One day it would be revealed to me. In the meantime, I had to get through each day as best I could.

Let me tell you something about deformity that I didn't know at that time, but know now: It teaches you a lot. It either ruins you completely and makes you useless to yourself and everybody else, or it makes you strong, so strong that you can overcome almost anything. *You* become the one who comforts the normal people around you, and it turns out you help them to bear your suffering. I know that's upside down, but that's the way it is. And that's one of the reasons I'm writing this book—not just to tell you about myself and all the miraculous things that have been done for me, but to tell you what it's like to be different, deformed, set apart, closed away, turned into yourself by the fact of your difference.

I want to tell you about this now because the world is changing very fast. People with twisted bodies and minds, once locked up in institutions and sometimes in attics, are coming out of their hiding places. It is now the law that public education must be provided for all children with handicaps. More and more of them will be educated in regular public schools right beside normal children; that's the law too. All of us will have to look at sights that are not always pleasant.

You won't be able to shield your children from knowing about children with frightening defects. These different children—deaf, blind, or with misshapen bodies or deformed faces—may be sitting next to your child in third grade. Or at the next table in the school cafeteria. Or they'll be rolling their wheelchairs or tapping their canes up the ramps to the high-school library or to the bleachers of the football field.

They're going to be among you, right in your midst, hundreds and thousands of them. And for your sake, as well as theirs, it's going to be important for you to know what it's like for them.

# ∞ 2 ∞

# *How It All Started*

I was born on December 31, 1955, in Curry Clinic in Chattanooga, Tennessee, the fourth and last child of Sarah and Edward Fox. My parents were both from farm families that had settled in southern Tennessee many generations ago. Daddy was the oldest of ten children (eight boys and two girls), and Mama was the fifth of eight children (five boys and three girls). They first met when they were working on a farm at harvest time. My sister Joyce was sixteen, my brother Bobby fourteen, and my brother Ralph twelve when I was born. All were healthy, normal, and good-looking.

My mother, who was thirty-seven then, had a terrible pregnancy with me—far different from her other pregnancies, which were very easy. With me she was sick and miserable almost the entire time. Dr. James McKinny, our family doctor, made her stay in bed a great deal, and Joyce took over the housework and cooking. Joyce was a good cook, but she had a hard time keeping the boys filled up, because Mama was so queasy she couldn't stand the smell of cooking. Everybody was eager for the baby to be born and for the family to get back to normal.

I don't think Mama realized she had anything serious to worry about until a little while after I was born. (I weighed in at six pounds and three ounces.) Then, suddenly, the doctors and nurses were too busy to show her the new baby.

"When do I see my baby?" she asked Dr. McKinny.

"The nurses are very busy," he told her. "They'll bring her to you as soon as they can."

A little later she repeated the question to a nurse: "You know, I haven't seen my baby. When are you bringing her?"

The nurse hesitated, then said, "I guess they're cleaning her up now."

Meantime, the doctor broke the bad news to Daddy. "I'm sorry, Mr. Fox," he said, "but something's gone wrong. It must have happened during your wife's pregnancy. Your baby is not normal."

The doctor said he couldn't give a name to what was wrong with me—it wasn't any well-known, recognizable condition or disease. Something must have happened to damage me while I was developing as a fetus.

Daddy saw me when I was a few hours old, and even to this day my heart goes out to him for the shock and pity he must have felt. I've never seen my own pictures at birth, and I don't think I ever want to see them. But I've heard and read the descriptions of what I looked like.

Poor Daddy! What he saw was a baby girl without a right hand and pretty much without a face. My eyes, as I've already told you, were off to the side. At birth the left eyelid didn't open at all. There was no nose—only a hole where the nose should have been. I had no upper lip—just flaps of loose tissue that resulted from a double harelip.

Inside my head there was a double cleft palate, with the cleft extending deep into the bony structure of my skull.

Someone once said that I looked as if a firecracker had gone off in the middle of my face—and that's about the best way to describe me. The hospital records reported that I had fifty-nine abnormalities of the face, skull, palate, and limbs, including my missing right hand and bands of tight fibrous tissue around my left leg and right thigh.

Years later, a woman remarked about me, "I'll never understand why the doctor let that child live."

A nurse who was in the delivery room when I was born happened to overhear her. "How dare you say such a thing!" the nurse rebuked that woman. "Who knows what that child will do in her lifetime?"

I am grateful for what the nurse said, because I have never been sorry that I lived. I am happy, really happy, that I lived! Even though I have had to struggle and suffer to obtain what almost everyone else takes for granted—a face—I now believe my survival was an important part of God's plan for helping other afflicted children. *I truly believe that.*

Maybe in a more sophisticated hospital in a bigger city the doctors and nurses might have turned their backs on me until I quietly died. I've heard that that happens to a lot of seriously deformed babies. I can only thank God that Dr. McKinny did everything he could for me. I was rushed by ambulance to Children's Hospital (which is now part of Baroness Erlanger Hospital) in Chattanooga, where they had better facilities. I was put in an incubator and fed with a medicine dropper, since my mouth was too messed up to suck on a bottle or a breast.

In the first week of my life, Dr. Barnwell began a series of surgical corrections of my abnormalities that are continuing up to this very day. In my twenty-two years I have had fifty-eight operations.

In that first operation he repaired the cleft in the right side of my upper lip as a first step in closing my upper lip. He noted on the hospital record, "This is one of the most severe and extensive cleft and facial deformities I have ever seen."

When I was five weeks old, he closed up the left side of the lip so that at least I had something that could pass for a mouth. At four months, he worked on my left eyelid so it would open. (A famous eye specialist told me many years later that if that corrective operation had not been performed at that time, I would have completely lost sight in that eye.) When I was ten months old, he released the constricting bands on my thigh and leg.

When I was born, all that Dr. McKinny told Mama was that I had a harelip that could easily be corrected. "She's been taken to Children's Hospital to have it worked on," he explained. Mama's blood pressure had shot way up after I was born, and the doctor wanted to keep her as calm as possible. Then, gradually over the next few weeks, Daddy and the doctor told Mama, bit by bit, until finally she knew it all—she knew the worst.

I was about three weeks old when they first let her come to see me in my incubator. She was allowed to hold me for a few minutes, and as the tears rolled down her face, she prayed to God and asked Him to reveal to her the reason why He had made me that way and why He had given her such a child.

We live in a part of the country where religion is very important—it's in the center of everyone's life every day. We see God as a living presence. We believe we are tested by adversity, saved by our faith, born again. But right from the start, Mama and Daddy knew that God would never punish an innocent child by giving her a ruined face. Neither would He punish the parents' sins by harming their child. Until His purpose was revealed to them, Mama and Daddy knew that it was God's will that they accept this strange, afflicted child He had sent them. There had to be a reason why God had made their child this way. They prayed to know it.

I stayed in the incubator for a month and in the hospital for four months. In between those earliest operations on my mouth, my parents brought me home on weekends. You're probably wondering how my brothers and sister reacted to the sight of me. Well, don't forget that children are very accepting. Joyce, Bobby, and Ralph hadn't had a baby around the house that they remembered, so I don't think they found anything especially upsetting about the way I looked. Later, by the time they'd heard people talk and had begun to realize how different and strange I was, they'd gotten used to me. I was their baby sister, Debbie.

I honestly can't ever remember a time when the bigger children were unkind or mean to me. They played with me, protected me, spoiled me some—and spent most of their time leading their own lives, teenagers busy with school or work.

Mama was the one who had the hard time. She was very nervous about feeding me with the medicine dropper, and sometimes cried with frustration when the dropper wouldn't work or I fussed for more milk, and then more, because it was a slow, tiresome business getting it all into me. And I was a very hungry baby. Maybe that was how

I got so attached to Daddy. Many times he took over the feeding when Mama was just too exhausted to keep at it. He held me in his strong arms and worked away with that tiny medicine dropper, dripping the milk into me through the bad excuse for a mouth that Dr. Barnwell was doing his best to patch together.


Naturally, I don't remember much about my early years. I'm told that I was a good baby, that my hair came in long and blond, that I never had to be spanked, that I was a great favorite of the nurses and doctors every time I went to the hospital for another operation on my mouth and lips. The nurses wheeled me around in a little go-cart and fussed over me. Daddy took me for rides in the car every day when he got home from work. I loved those rides and could hardly wait for him to get home so I could run out to the car and climb into the back seat.

Joyce got married when I was three and had two babies, one born in 1960 and the other in 1962. I thought those babies were better than dolls. They were prettier than my dolls, anyway. They were warm and soft when I held them, and they made lovely gurgling noises. Except for my hand, it never occurred to me that I was different in any way from Mike and Jimmy, Joyce's beautiful babies. (Remember, I had not yet seen my face.) My brother Bobby got married to Janice, who had lovely red hair. Janice often sang to me and rocked me and changed me. When I was a little bigger, she walked me down to the brook and I saw the children there swimming or fishing. The only thing that really bothered me was not being able to go down to the brook by myself and play with the children.

Our house is at the end of the road, almost at the foot of Soddy Mountain, with the endless freight trains of the Southern Railroad passing at the base of the mountain.

You've heard about that lonesome whistle of the Chattanooga choo-choo. Well, those blues in the night were my earliest lullabies. Soddy and the next town, Daisy, are small rural communities in Hamilton County, just north and east of Chattanooga.

When I got to be about four or five, I began to notice that groups of children were standing near our house in the morning, waiting for the school bus. In the afternoon I saw them come off the bus, laughing and chattering, their arms full of books. I asked Mama when I would be going to school on the bus. She told me I wasn't big enough. That was true when I first asked.

But then I got to be nearly six, and I knew I was getting old enough to go to school.

"Mama, when do I start school?" I asked.

"Oh, pretty soon now, I think."

That wasn't definite enough for me. "How soon? Next week? Do you think next week?" I couldn't wait to run down the path with my notebook under my arm and get on the big yellow school bus with the other children. Once I was on the bus, they'd talk to me and play with me. I'd be part of them. I'd no longer be alone. "Next week, Mama, please? Please?"

"Not next week, but pretty soon."

I can't tell you how many times Mama and I repeated this conversation.

After a while I began to get worried. I was as old as some of the children who got on the bus—why couldn't I be there with them? I looked at my little stump of a hand. I hated the sight of it. Maybe If I put on a glove, no one would notice. Maybe Mama was waiting for the cold weather; then I could wear a mitten and kind of forget to take it off.

I asked Mama about that. "If I wear a mitten do you think they'll let me on the bus?"

Mama looked at me in a funny way, and I could see she was crying a little. "No, Debbie, the mitten won't help. But pretty soon a teacher is coming to teach you here at home."

# ～ 3 ～

# *To School—at Home*

I remember how excited I was that first morning of school. I got up very early and put on my best dress—navy blue with white cuffs and white piping down the front. Mama helped me into my clothes, as usual—I couldn't manage buttons and hooks with only one hand. She brushed my hair, which hung long and blond over my shoulders. Joyce had given me a pencil box. Daddy had bought me a notebook just like the ones I'd seen the other children carry. Finally a car drove up, and I peeked through the curtains of the living-room window to get a first look at my teacher.

She had dark brown hair fluffed up high, a kind face, and warm brown eyes. I was standing right at the door when Mama opened it, my notebook tucked under my right arm and my pencil box in my left hand. The teacher walked up the three steps to our porch. When she saw me, she suddenly stood still. I thought for a minute she wasn't going to come any closer and I began to feel cold and frightened. Then she smiled and kept walking toward me and said, "You must be Debbie." She said she was Mrs.

Madeline Apple from the Hamilton County Board of Education.

Mrs. Apple sat down in our living room, and she and Mama talked a lot about things I didn't quite understand. I was disappointed when she didn't start giving me lessons, because I had thought I'd get to use my notebook and pencils right away. After a while Mrs. Apple got ready to leave, but first she said to me, "Debbie, I'm going to arrange for a teacher to start working with you later in the week. Will you like that?"

"Yes, ma'am," I said.

I was very disappointed. I liked Mrs. Apple and I wanted her to be my teacher.

Later, when Mrs. Apple got to be a very important person in my life, I found out what really happened that first day. When I got to be six, Dr. Barnwell applied to the Board of Education for a visiting teacher for me. The assignment was given to Mrs. Apple, who was then in her third year as a home teacher of the handicapped. Her information sheet about me said only that I had a cleft palate. She expected to see a child with just a minor defect. That was why she wasn't at all prepared for the sight of my extreme deformity. In fact, my horrible face stopped her in her tracks as she came up to the porch. She was so shocked that she was ready to turn and run. Only the stricken look in my eyes forced her to go on. She made herself walk into the house. She talked to Mama and me and then went straight home, canceling all her other appointments for the day. She could hardly see to drive home, the tears were running down her face so hard. She was just shaken apart.

She was still terribly upset that afternoon when she telephoned her boss, Edward Fitch, the assistant superintendent for pupil personnel services, and described the frightening sight she'd seen at the Fox house in Soddy.

"I don't think I can do it," Mrs. Apple told him. "I don't think I can face that child."

The assistant superintendent told her she didn't have to; he'd give the assignment to another teacher. But the next day Mrs. Apple went to see him.

"I've thought some more about that child in Soddy," she said. "I spent all last night thinking about her. I hardly slept. I can't get her out of my mind. There's something special about her. She doesn't have anything to smile with, but she's smiling from inside. I think she's been sent to me for a reason."

And so Mrs. Apple began coming to our house three times a week. Teaching me, I'm sure, was the worst struggle of her life. I caught on quickly in reading. That was easy. I had to hold the book off to the side and look at it with just one eye—usually my right eye, which is stronger than my left eye—but that didn't slow me down. I read right off.

But writing was a battle—a battle that seemed to last forever. You see, I'm naturally right-handed—but I don't have a right hand. So I had to learn to write with my left hand. I think you'd have trouble, too, if you were forced to write with the wrong hand. All I know is that I just couldn't get that left hand to do what I wanted it to do. The letters sprawled and tumbled.

Mrs. Apple brought me huge sheets of brown wrapping paper and a giant fat pencil. I practiced my letters and numbers. Oh, how I practiced them! After I had my hand under control, there was the problem of coordinating my eye and my hand. That trick was very slow to develop. Since I used only one eye at a time, I had no depth vision. That meant everything looked flat and in the same plane. And, of course, I had to twist my head around to look at my left-handed writing with my good right eye. It was *some* struggle, I tell you.

Mama set up a desk for me in the living room and a shelf for my books. Every day I sat there, with Mama leaning over me—helping me with my letters, listening to me read, going over the homework that Mrs. Apple left for me to do. On the days when Mrs. Apple came, I was waiting for her on the porch. If I'd read ahead of my lessons on my own, she was pleased and praised me.

"That's very good, Debbie," she'd say. "You're really trying hard. I can see that."

She was kind, but strict; she never let me get away with anything. If I'd been lazy and hadn't finished the work, she'd scold me. "Debbie," she'd say, speaking slowly and distinctly, "I wonder if you're really trying. I don't think you are. I know you can do better."

I'd do better most of the time. But once in a while I'd slip, and then Mrs. Apple would catch me up and make me toe the line. I wish I could make clear to you how badly I wanted to learn. Mrs. Apple once said that I was like a sponge, soaking up everything around me. Maybe I was so eager to read because it was a kind of miracle to me. After all the time alone, by myself, shut away from the other children, suddenly I could pick up a book of Bible stories or *Sleeping Beauty* or the Dr. Seuss books with all the funny animals and be somewhere else—with people who didn't know I was a sad little girl, looking out into the world from a closed room. So I absorbed everything Mrs. Apple brought me. But it wasn't always easy.

Sometimes there were tears, almost tantrums, as my written words wandered uphill and down on the paper and my awkward hand formed shapeless blobs for letters. It was a real worrisome time for all of us. Once, I remember, I just couldn't get the lesson right and I'd used up all the wrapping paper Mrs. Apple had left for me. "I can't do it, I can't!" I wailed, all frustrated and angry.

Daddy had just come home from work. He had a job

with a big machinery company where he spray-painted the heavy equipment. At the end of the day he was tired right down to his bones. But that day, as he always did, he said, "Come on, Debbie, let's take our ride."

I climbed into the car with him, and he talked gently and soothingly to me as we drove deep into the green hills where the trees met above the car. We were in a dark tunnel of trees, and the leaves whispered over our heads. Then we went around a curve, and Daddy pointed straight ahead. "Look, baby," he said. "See the light there?"

There was just a small dot of light off in the distance where the road came out of the trees.

"That's called the light at the end of the tunnel, baby, and I want you to remember it because it means hope. It's a promise from God that you will come out of the dark."

We drove home then, and Daddy handed me a big sheet of the Chattanooga newspaper. I spread it out on my desk, and with my fat pencil I began writing my lesson across the newsprint. I tried harder than I ever had before, and after a while the words came out just the way I wanted them to. I knew Mrs. Apple would be very proud of me the next day. *Light at the end of the tunnel.* I liked that idea a lot. *Light at the end of the tunnel!*

# ∽ 4 ∽

# *The Telephone Was My Teacher*

When I got to the third grade, something wonderful happened: A telephone hookup was installed in our living room which connected me to my third-grade class at Soddy Elementary School. I could hear Miss Anderson, the teacher, and the children and everything that went on in the classroom. When it was my turn to recite or I wanted to ask or answer a question, I just had to push a button on the red box in our living room and the teacher and the boys and girls several miles away could hear my voice. I stood up to recite the Pledge of Allegiance with the other children. At lunchtime I could chat back and forth with my classmates.

At the time I took the telephone hookup for granted. I didn't realize there was anything special about it. Later I found out it was *very* special. To explain how it happened, I have to go back and tell you a little about Mrs. Apple, who has played such a big part in my life. Mrs. Apple grew up in Dayton, Tennessee, about ten miles farther down the road that leads from Chattanooga to Soddy. You may

think you've never heard of Dayton, but I bet you have—
it's the place where that celebrated "monkey trial" was
held back in the 1920s, the one in which William Jennings
Bryan, the golden orator, and Clarence Darrow, the famous
lawyer, argued about the teaching of evolution in the public schools of Dayton. Even to this day, there are some
older men around our way who were at that trial and still
talk about it.

Anyway, in the 1950s Mrs. Apple was teaching sixth
grade at White Oak School in Hamilton County. She was
married to Arlie Apple, a tall, thin, very kind man who was
an engineer with TVA. The Apples had no children of
their own, and maybe that's one of the reasons why, when
the time came, she took to me so strongly.

One day, after she'd been teaching about ten years, Dr.
Sam McConnell, the Superintendent of Schools, sent word
that he wanted to see her. I guess when a teacher is sent
for by the superintendent, she gets as nervous as a kid who's
sent for by the principal. Mrs. Apple remembers that she
was real worried, almost shaking, when she went in to see
Dr. McConnell.

He's a tall, lanky man, built the way Abe Lincoln must
have been, and it seems he didn't want to scold her at all—
he wanted to offer her a new job.

"We're opening up a program for homebound children,"
he explained. "Teachers will be going into the homes of
children who are too crippled or sick to go to school. The
teachers will give them their lessons in their own living
rooms or even in their own beds."

That was a very advanced thing to be happening in a
rural school district at that time, but Chattanooga
and the little towns around it, I learned later, have always
been out front in education.

"I want you to teach in that program," Dr. McConnell
said.

Mrs. Apple wasn't sure she wanted to leave her sixth grade, but after thinking it over for a week, she finally told Dr. McConnell, "I'll try it for a year. If it doesn't work out, I'd like to go back to my sixth grade." The way it happened, she stayed twenty years with the homebound and retired in 1977 as coordinator for special education.

The children she taught had cleft palates or brain tumors or legs twisted by polio, or they were in wheelchairs because of cerebral palsy or muscular dystrophy. While I wasn't as sick as some of those children, in other ways I was worse off, because it was so hard for people to look at me.

After Mrs. Apple began coming out to Soddy to teach me, she kept telling the people at the Board of Education about the child who hardly had a face. "She's really remarkable, that little girl," she told them. "She barely looks human—but she's the best pupil I ever had. Nothing stops her from learning."

They found it hard to believe what she was telling them. One day Dr. McConnell decided to come and see for himself. I didn't realize at the time how unusual it was for a school superintendent to go see one particular child at home; I took his visit, as I took almost everything in my life, as normal and everyday. How was I to know otherwise?

So Dr. McConnell came to our little house on Soddy Creek. I saw so few people then that I was always interested in someone new. I smiled at him and recited a little poem I'd memorized and read a few lines from my reading book. He didn't say very much, except to tell me that I was doing fine and to tell Mama that he admired the way she was working with me and helping me with my lessons.

Later, in his office, Dr. McConnell kept shaking his head and saying to the people on his staff, "I'd never have the imaginative power to build up for what I saw in that house. There was such eagerness in the little girl's eyes—in those strange, distorted eyes. And when I saw the compassion and feeling in her mother's face, I knew we had to do everything we could for that child."

Dr. McConnell was aware that Dr. Barnwell was working hard to straighten out my face. That was why he once said, "It's a challenge for us to keep up with Dr. Barnwell. If he believes she has a future, we have to go the extra mile to reach the goal he's set."

I didn't know about any "extra mile." All I knew was that I was doing my best to learn and that Mama, Daddy, and Mrs. Apple were pleased with me. And when I reached third grade, Mrs. Apple began with the first of what I've come to call her "schemings." She's the kind of person who wakes up in the middle of the night with an idea. And when she does, there's nothing that can shake that idea out of her head—nothing in the whole world.

To get a little ahead of my story, I think I was only about thirteen or so when she first said, "You know, Debbie, you should start remembering things now, so you'll have them in mind when you write your book."

Write my book? Me? A little country girl from Tennessee who wasn't even in high school yet? It was crazy! Only important authors wrote books. I couldn't even begin to take the idea seriously—not until years and years later. But way back then Mrs. Apple had in mind this book you're reading now. She pushed and fussed and even nagged about it—until it happened.

Well, when I reached third grade, she got this idea about the telephone hookup for me. I guess she'd read about it in some teachers' journal. Hamilton County had never had telephones for the homebound. That didn't bother Mrs. Apple. She made phone calls and wrote letters to find out how it was done. She told Dr. McConnell what she wanted, and she kept on telling him. She never let up about that telephone—not until it happened.

Looking back now, I realize that in addition to bringing me closer to my classmates, the phone set two important patterns in my life. First, it made it possible for my shortcomings to help others. Because I needed a phone, other handicapped kids got phones too. Much later, because I needed a face, others were able to have their abnormal faces made to look more nearly normal. I'll talk later about the miracle of helping others that has given so much meaning to my life.

The second thing the phone did was to give me a specific goal to work toward. I'll tell you right now about that goal.

I guess the Board of Education wanted to show the taxpayers how much good was being done with their money. So when the telephone was put in, some of the school officials came to have their picture taken with me for the local newspaper. The photographer lined me up with Mama and Daddy and Mrs. Apple and Dr. McConnell and maybe some others.

There was some whispering among the grown-ups. Then Mama took me gently by the shoulders and swung me around so that my face was away from the camera. We didn't say anything to each other. But our eyes met, and her eyes said to me, "Debbie, I'm sorry," and my eyes said, "It's all right, Mama, it's all right."

By then, of course, I knew about my face and the reason for not showing it. I think it was from that moment, on the porch of my house, when all the others looked forward and I looked backward, that I made up my mind that I was going to get my face fixed, no matter what. *I wasn't going to let anything stop me.* Not my fear of surgery, not pain, not money, not anything. I wanted to look like other people. I didn't want to turn my back anymore. I wanted to face *forward*—right into the camera, right into people's faces. And I knew that God was with me and would give me strength in my fierce longing to be whole.

The school telephone did something else for me: It began to give me friends. I got to learn the names of the children in my class. Gradually, from their voices and from the things I hard them say, I began to tell them apart. At lunchtime, some of them talked to me, asked me what I was doing after school or told me a joke or a funny thing that had happened. Two of the girls who were especially friendly were Karen McRee and Kathy Smith. After we got to know each other on the phone, they came to see me sometimes after school. I suppose they'd been warned about how I looked, but they never showed a thing on their faces. Both Karen and Kathy were very pretty girls, and I could tell from the conversations I heard that they were very popular with the other kids. What was more, Kathy was the class brain—the smartest of them all. I was real proud to think that popular girls like Karen and Kathy wanted to telephone me in the evening and sometimes come visit with me.

Miss Anderson came to our house a number of times to help me with my arithmetic or get me ready for big tests. I was always excited when she came to visit—any new person who entered my small world was important in my life. Several times I asked Mama if I could go to school to meet the boys and girls whose voices were becoming familiar to me. Mama always changed the subject very quickly. She knew, and I guess I knew too, that I wasn't fit to look at. I wanted so badly to meet the children and see what my classroom looked like.

Sometimes I'd close my eyes and try to imagine the classroom. In my mind it was very big and the blackboards were very black. I'd imagine the flag in the corner, the pictures of the Presidents on the wall, the cut-out turkeys and Santas and Easter rabbits hung against the windows at holiday time. Later, when I really went to school, I had to laugh at myself because the classroom I'd imagined was nearly three times as big as a real-life classroom. But I was right about the pictures and the flag.

It made Mama so happy to hear me on the phone, laughing and giggling and saying to Karen, "Tell me about the party! Who was there?" If I couldn't go to the parties and gatherings of my classmates, at least I could hear about them. Somehow that made me a part of what was going on.

A girl about my age named Darla Johnson lived fairly near me. Every so often she'd come by to say hello. And sometimes Karen or Kathy would come to my house to pick up my homework and take it to school for Miss Anderson to correct.

Now I wasn't as completely shut in as I had been. And about this time, I began to have my little nephews, Mike and Jimmy, for playmates. Joyce took a job and every morning she dropped the boys off at our house so Mama could look after them. At first the boys were in a playpen in the dining room, which opens right off the living room. I'd be at my desk, with my books and workbooks around me and the telephone in easy reach. My teacher would be talking about arithmetic or geography, and Mike and Jimmy would listen, just the way I did. Mama had to give them pieces of paper and pencils and baby books, so they could "go to school" too. Then, when they got tired of playing school, Mama had to think of ways to amuse them so they'd be quiet and not interrupt me. Mama was very strict about not letting them interfere with my school time.

What really kept me going at this time—the shut-in time of my life—was my own faith and the faith of those around me that I truly had a future. Why would Mrs. Apple bother to come all the way out to our house if she didn't believe that someday I was going to lead a real life and have to know how to add and subtract and remember the name of the capital of Oregon and who was President after George Washington?

If she believed in me, I had to believe in myself. Much later, Mrs. Apple told me that she pushed and drove me so hard because she knew, even then, that I had a future. Some of the children she taught were too sick to grow up. They were given lessons to make their lives *look* normal, but there was no real hope for them. Any number of times her phone rang in the middle of the night and a heartbroken mother or father said to her, "Please don't come to the house tomorrow—we just lost Eddie—he's gone." Or Paula. Or Gay. Or whoever.

"Debbie, you've got a really good mind," Mrs. Apple would tell me. "Now you've got to work hard and make something of yourself. When Dr. Barnwell finishes with your face, you'll be able to do whatever you want with your life."

Mrs. Apple believed in my future, Dr. McConnell believed, Mama and Daddy believed. I had to believe too. I have to tell you that I had my moments of doubt. I tried not to look at myself in the mirror. But sometimes when I saw my face reflected in a window, the tears would start again and I'd wonder: Is it really possible to fix that face? Will it really happen?

## ∽ 5 ∾

## *They Believed in Me*

Two other people who believed were my dentist, Dr. James Phillips, and my plastic surgeon, Dr. Howard Barnwell. I usually went to see Dr. Phillips at seven thirty on Saturday mornings. At the time I didn't think that was an odd hour to see a dentist. Now I realize that Dr. Phillips set those early Saturday appointments so I wouldn't be uncomfortable about meeting other people in his waiting room. My jaw was so incomplete that he had to do a lot of difficult work to get my teeth right as they grew in. I had braces, fillings, and caps on my back teeth. I was always nervous on those visits, even though Dr. Phillips did his best not to hurt me.

When I got out of his chair, he always gave me a dime to buy ice cream at the soda fountain downstairs in his building. Just recently he asked me, "Debbie, do you remember the time I forgot to give you the dime when you were little and you tugged on my coat and complained, 'Where's my dime?'"

I'm afraid I don't remember, but I want to tell Dr.

Phillips, right now, how grateful I am for all those Saturday mornings when he could have slept late or played golf, but instead got up early to work on my teeth.

I found out later there was something else I should be grateful about: Dr. Phillips never charged us for the work he did on my teeth. He was that kind of man, and besides, my case was so extreme that he looked on it as research.

Dr. Barnwell was a tall, outwardly gruff man who had his practice in an office building in downtown Chattanooga. Daddy usually drove us—it took about half an hour from our house. In the early years, when I still didn't know about my face, I walked boldly into the elevator and stared back at anyone who stared at me. Later, when I knew what I looked like, I cowered in a back corner of the elevator and tried to hide myself. Sometimes I turned my back, as I had for the photographer, and looked at the wall.

Usually the doctor's waiting room was full of children and grown-ups—some with casts on their arms or legs or patches on their eyes. I'd read a storybook while I was waiting. The receptionist always said, "Ah, here's my little friend again." And Dr. Barnwell gave me lollipops or candy.

I liked the way he looked at me. Now I realize he was looking beyond my horrible face to the new face he was planning for me. By the time I was twelve, he had already performed nearly three dozen operations. He had fixed my soft palate, worked on the clefts in my hard palate, tried to make that balky left eyelid behave, done some corrections of my nose, started to reconstruct my right nostril, X-rayed my right arm and my skull. He did the small corrections in his office. For the bigger ones, I had to go to the hospital.

Even though I was in the hospital for only a few days at a time and usually healed very quickly, I hated the hospital. I hated the operations, the pain, the bandages, the needles that someone was always sticking into me. Most of all, I hated it when Mama and Daddy had to leave and I was alone with only the nurses to comfort me when I cried. When I was very little, I took my teddy bear to the hospital with me. Later I had a silly little yellow rubber elephant that I wouldn't go to the hospital without. The elephant wasn't cuddly or furry, but it cheered me up and I loved it.

Sometimes Mrs. Apple went with us to Dr. Barnwell's. Then she and Mama and Daddy would talk to him about his plans for me. He had taken stacks of photographs that showed every stage of correction. He'd also made plaster models as he went along. In addition, he had drawings and plaster renditions of my face as he hoped it would look when he was finished. He thought the work would be completed when I was about sixteen.

I never saw those sketches, but I'd listen and try to understand what the grown-ups were saying. I'd try not to cry when they told me I'd have to go back to the hospital for another operation, but sometimes I just couldn't help it. The thought of the operating room and being left alone in my bed at night would get to me. Even though I wanted more than anything to have my face fixed, I could feel the tears sliding down my cheeks.

I remember once when that happened, Dr. Barnwell came over to where I was sitting and stood looking down at me. He was a big bear of a man with huge hands, so big you wouldn't think they'd be able to do all the delicate cutting and tying needed in plastic surgery. Dr. Barnwell was famous all through the Southeast for his reconstructive work. They called one ward at Children's Hospital "Barnwell's Babies," because that's where he worked on afflicted children, and often he didn't charge their families.

Plastic surgeons usually weren't held in such high regard at that time. Some of the other surgeons, I heard later, called them "skin stretchers." I guess doctors, as well other people, didn't know what wonders they did for children with birth defects or for children and adults who had been disfigured by accidents and malignancies.

One little boy who'd put a live electric wire in his mouth was in and out of Children's Hospital quite a lot when I was there. You wouldn't believe what Dr. Barnwell did for that child's burned-out mouth! He repaired it and worked on it until it was a perfect little rosebud.

Well, that day when I was crying, Dr. Barnwell wiped away my tears with his big, strong hand and said, "Debbie, you're going to have the face you always wanted! And you know what? When you get married, I'm going to take you down the aisle myself."

That was what I wanted to hear, because I longed so desperately to look like other children. More than anything, I wanted to look like my sister, Joyce. She has blond hair, blue eyes, straight eyebrows, fine features, pretty coloring, a perfect nose, and a perfect cupid's-bow mouth. At that time, I had blond hair and blue-gray eyes and that was all. The rest of my face was as far away from Joyce's as anyone could ever imagine. Sometimes I'd catch myself staring at her. I wanted to memorize her face, because I wanted to have one exactly like it.

Joyce did a lot of things for me, and one of the best was to try to make me a little more independent. At home, I was Mama and Daddy's baby; being the youngest by so much, I guess that they would have babied me even without my problems. Joyce never did.

Once when I was with her I needed to have my hair brushed and combed. Mama had always done it for me. But Joyce just handed me the brush and comb and went on with what she was doing. I was about to say, "Please do it for me," but suddenly I realized my big sister was treating me like any other ten-year-old. It came to me then that if I wanted to be like other children, I had to begin to *act* like them. So I picked up the comb with my left hand and worked it awkwardly through the tangles in my hair. Then I used the brush just the way Mama did, until my hair was silky and fluffy.

I don't think Joyce knew how much she did for me that day.

# 6

## Into the Valley

It was Mama's sister, my aunt Neal Howard, from whom
I first learned the terrible news of Dr. Barnwell's death.
The time was just before Christmas in 1967. I was almost
twelve years old. Aunt Neal rushed into the house with
the Chattanooga paper in her hands. "Debbie, look, can
you believe this?" she said. On the front page was a story
about the automobile crash on the road from Atlanta that
had killed Dr. Barnwell and injured his son. I grabbed
the paper from Aunt Neal's hand and read about the accident. I couldn't believe it. "Oh no! Oh no!" I kept saying
as I read.

It wasn't possible that Dr. Barnwell was dead. That huge,
strong man with the huge, sensitive hands! It couldn't be!
But it was. There was his picture, real as life, and the
terrible words telling how he died on the way home from
visiting his brother. I could feel myself shaking inside. I
started sobbing.

I was wet all over with tears when Mama came home.
When she heard the news, she began to cry, too. Daddy got

back from work soon after that. Mama and I couldn't talk. We just passed the newspaper to him. I'd never seen a man cry before. But there was Daddy, crying just as hard as Mama and me. We were crying for Dr. Barnwell, a good man, who should never have died that way. And we were crying for ourselves, too. What was going to happen to us now?

All our hopes had been pinned on Dr. Barnwell. He had worked on me ever since I was born. I trusted him and believed in him. And he believed in me and my future. He really believed that I *had* a future. And because Dr. Barnwell believed he could fix my face, Mama and Daddy had been able to hope that someday our nightmare would be over and I would have a normal appearance. Then we'd be a normal family again.

Now it was all over, our hope all gone. I don't remember what we said to each other that day. I don't think we said very much. We kind of huddled together and held on to each other. I think Daddy was the worst off of all of us. He kept shaking his head and saying, "I don't know, I don't know," as if he were just sort of giving up. It was pitiful to see the state he was in. Mama somehow got dinner on the table, but we didn't eat much. The phone kept ringing as word got around and people called to find out if we'd heard. Yes, we'd heard. We'd heard the worst.

Mrs. Apple would have rushed out to be with us, but she was in the hospital. A car had swung into her station wagon as she made a left turn near the Soddy-Daisy High School. She'd been badly shaken up and had to spend some time in the hospital to recover. We'd been through a few awful days when we worried whether her injuries were serious. I can't tell you how relieved we felt when we found out they weren't. But now she couldn't help us, and we had to handle our grief alone. "Daddy, what's going to happen to me?" I asked.

Daddy put his arm around me and tried to comfort me. "We'll get you fixed, baby," he promised. "We'll find another doctor. There has to be another doctor."

But another doctor wouldn't be the same thing. He wouldn't know me and I wouldn't know him. I could never believe in another doctor the way I believed in Dr. Barnwell.

I felt as if there must be a plot against me. My teacher in the hospital, my doctor dead. I felt lost and abandoned. Mama tried to put aside her own sadness to help Daddy comfort me. "It's going to be all right, Debbie," she said. "You have to have faith. The Lord will help us."

I wasn't so sure. I felt my faith at a low ebb. Maybe God really didn't want me to be all right. Maybe His plan was not to give me a good face after all.

When I felt my faith waver, I became frightened, really frightened. To lose my doctor was a terrible thing, but I could accept it. I had to accept it. To lose my belief in God's goodness—that was something else. I couldn't accept that. I couldn't even let myself think about it. I couldn't live if I didn't believe in God.

I've already told you that in my family, our faith is just about the most basic thing in our lives. We belong to the Church of God, a church that has many, many followers in our part of the country. Maybe this is a good time to tell you a little about it, because then you'll understand why faith is as natural as breathing for us.

The Church of God began about a century ago among the country people along the Tennessee–North Carolina border who wanted to be closer to the Scriptures and to God than they felt they could be in the regular Protestant churches. So they set up revival meetings and started little churches in the poor mountain villages.

People came to the revivals to praise God and to be healed. They felt close to God and to Jesus; they sang and shouted their praises. Some of them spoke in strange tongues when the Holy Ghost came into their hearts and blessed them. There are Churches of God now in all parts of the United States and missions all over the world.

I wonder if I can make clear to you how important church is to people like us in small towns in the South. We can't even imagine *not* belonging to church and *not* believing in God. We go to pray, to study the Bible, to listen to the preacher, to seek healing for the sick, to give thanks for our blessings, to be born again. Church is also the center of our social lives—it's where we meet our friends and where the young people get together.

Members of the Church of God put aside a lot of frivolous things. We are plain people. We don't drink or smoke or swear or wear jewelry or use make-up or go to the movies. All these things are forbidden, because we must keep ourselves holy and clean, as commanded in the Scriptures. We believe that through faith in the blood of Christ we will be saved and the righteous will have eternal life.

At the time when Dr. Barnwell died, I believed in God and His goodness, but I didn't know all this about the Church of God, because I'd never yet gone to church. I'd wanted to go for a long time. Mama had always put me off. When I was very little, I couldn't understand why Mama always said, "No, not this Sunday, Debbie, but soon, maybe after Easter," or another time, when I begged again, "Maybe after Christmas," or "Next year," or just "Soon." Not knowing that she was trying to protect me from hurt and unkindness, I pestered her and pestered her about church. It became almost an obsession with me. Each Sunday I would start my begging all over again.

After I knew about my face, I was more docile about being left home. Eager as I was to go to church and say my prayers to Jesus with the others, I was afraid to show myself. I was old enough to fear stares and cutting words. To hear them in God's house would be unbearable. So I was torn—part of me tugging me to church, and part of me pulling me back to the safety of my own home.

Now, with Dr. Barnwell dead, I was ready to forget my face. I didn't care how people reacted. I needed to go to church, to pray for Dr. Barnwell and for myself. Mama was adamant. I had to stay home.

Somehow we got through that dreadful time. Mrs. Apple recovered and started coming again to the house. I went on with my schoolwork. But I retreated more and more into myself. I felt alone and isolated. I wasn't sure whether or not I did have a future. For the first time I really doubted whether there would be a place for me in the world. I had always imagined that someday I would be pretty, find someone to love, get married like Joyce and have babies and lead a real life. Now I was no longer sure. I wasn't sure there was light at the end of the tunnel.

My mood got even darker as we searched among surgeons in Chattanooga and found there was none to carry on Dr. Barnwell's work. He had been the best in the whole area. We didn't know how to look anywhere else. None of the local doctors seemed to have any leads for us. We knew there were surely many surgeons in the big cities. But how would we find one? And if we did, how would we handle the travel and the expenses? Up to this time, Daddy's wages had taken good care of us. Blue Cross and Blue Shield had covered most of my hospital expenses. And when Dr. Barnwell remembered to send a bill, which wasn't often, it had been for an amount that Daddy could manage.

Besides, perhaps it was God's will, after all, that I should stay as I was and hide my face at home.

So I burrowed deep inside myself. When I wasn't following along with my class on the telephone hookup and doing my homework, I sat by myself, brooding. When spring came, I spent a lot of time outdoors. Ours is a neat white frame house with a green roof set back on a lawn, with a persimmon tree in the yard. Soddy Brook, just across the road, flows into Chickamauga Lake, which is part of the vast Tennessee Valley Authority. Back in 1940, when the great Chickamauga Dam was finished, President Franklin D. Roosevelt came to dedicate it, and many people still have framed in their living rooms the front page of the newspaper with President Roosevelt's picture.

The mountains just to the west of us that cut off the sun early on winter afternoons are part of Walden Ridge, which stretches northward far into Virginia and south down into Georgia. There is logging and coal mining on the other side of our mountain, and I would often see the huge trucks go by loaded with coal or logs. The coal goes to a steam plant that generates electricity for TVA, and the logs go by barge to a paper factory in Bowater, Tennessee.

From my window I watched the trucks on the road, the children at the creek, the trees and flowers as the seasons changed, the dogs and cats and other small animals. And I began to write poetry.

I wrote poems about the things I saw around me in that period when I felt separated and alone and not sure whether I had that future. Now when I look at those poems I realize they're not very great as poetry. There was a lot I didn't know about rhyme and meter. But I'm really touched by these verses that were the outcry of a lonely little girl reaching to nature and the seasons and the beauty of the earth for the solace and the hope she needed to go on with her life.

Here is a poem I wrote at that time, called "Autumn."


### AUTUMN

> When leaves begin to fall
> And the corn has grown so tall

The pumpkins are big and yellow
They look like the moon when it's mellow.

Leaves are such beautiful things
When mother nature brings
The time of year
That gives the best of cheer.

Leaves are red, yellow, brown
Before frost brings them down.
In them children love to play.
Wind will soon blow them away.

We began to rake and burn
Getting ready for the return
From leaves to fallen snows.
And sleds are planted in rows.

Here's one about my beloved Soddy Creek called "The
Winding Creek."

## THE WINDING CREEK

I live beside a rocky creek
That winds and turns, but very meek.
In winter time it overflows,
In summer it hardly goes.

The water ripples over rocks
The sounds it makes seem to mock
All the unfamiliar sounds at night
Made by tumbling rocks.   What a fright!

Underneath these big waterholes
Little fishes are so bold,
Playing and swimming to and fro
As if they had no other place to go.

I love to play in the water holes,
To swim and play with fishing poles
From early morn till late at night
When the sun goes out of sight.

It's such a pleasure for a child
To play and have the ideas wild
Of swimming in this rocky place.
Yet soon our childhood pleasures do erase.

And here's our mountain:

### A Mountain

Have you ever viewed a mountain side
With trees overhanging huge and wide?
In spring its leaves are all shades of green,
It makes such a lovely scene.

I live just below this mountain.
The breeze is as cool as from a fountain.
It's so refreshing on our sweaty brow
That nature has planned somehow.

It has such beautiful flowers
Towering up toward the sky,
Created by a higher power
So all nature can magnify.

The streams that flow by its side
Are cool, refreshing, and wide.
It's so pleasant for little feet
To cool from the beating heat.

In fall its leaves are many colors
As old mother nature over it hovers—
From greens to orange, yellow, and brown
The leaves finally drop to the ground.

I told you that the Southern Railroad runs right near our house. Here's a poem I wrote about our railroad.

### RAILROAD

Have you ever lived by a railroad?
What a place for an abode.
It's really noisier than you think
When trying to study or think.

When I was a child that train would fascinate.
Its sounds I would try to demonstrate,
Watching it go sailing by
Like a huge bird in the sky.

Some of them have huge smoke stacks
Sending cinders down the tracks.
Others burn a form of oil
That forms a film on the soil.

It hauls all kinds of transports
From land to sea it exports
From dealers in United States to foreign lands
To replace orders in great demands.

The sounds are such a rumble and roar
It is indeed a chore
To listen to TV or talk on the phone
When it is traveling in your zone.

I hadn't glanced at these poems for many years until I gathered them together for this book. Now I'm astonished at them, at my powers of observation as a child of twelve, at the sense I had of life rushing by, of my grasp of life's realities.

And even though I had not yet been to church or Sunday school, I had read my Bible stories faithfully and I knew Jesus was watching over me.

## CRUCIFIXION

On Friday Jesus was crucified
So cruel men could be satisfied.
Up Calvary's hill, tortured and tried,
A drink of water He was denied.

When He traveled as far as He could,
They made Him carry His cross of wood.
So cruel they treated our blessed Saviour
Who had done no harm but favor.

On top of Golgotha He was nailed
And to be sure they hadn't failed
Placed spikes in His feet and hands
To hold secure their evil plans.

They hung Him there in the sun
In awful mockery and cruel fun.
"A drink of water," He cried,
But, of course, He was denied.

He then was placed in a tomb
Which was given by another man.
They thought Jesus was surely doomed
But there was yet another plan.

The third day He arose.
His body did not decompose,
But full of life did home return.
For this place we would yearn.

God gave His only son
That we be free and love
And accept all, reject none
At last—reach heaven above.

I may have thought I was forgotten at this time, but truly
I was not.

## 7

# *Hope in Atlanta*

One who never forgot was Madeline Apple. After all the trouble I'd given her, Madge wasn't about to let go. She was really determined in her conviction that I had a good mind and a lot of promise. And she just would not let me fall back.

"I could see that longing in your mind and in your heart," she told me later. "I knew you wanted to be someone. Somehow I knew I had to find a way to get you to a good doctor who could carry on Dr. Barnwell's work."

Why did she care so much? I can't tell you for sure. Maybe the answer's very simple—maybe God sent her to help me.

From her own doctor Mrs. Apple got the name of a plastic surgeon in Atlanta. She got in touch with him and found out that he would see me. Then she mentioned her plan to Mama and Daddy. They weren't sure what to do. Atlanta is three hours or more away by car. It's another world, a big, strange city. We were not used to traveling. There would be new people, new confrontations for me.

Maybe I would be exposed to cruelties. Maybe it would be better for me not to reach so far.

The four of us—Mama, Daddy, Mrs. Apple, and I—sat in the living room talking back and forth. Should we go to Atlanta or not? First we decided one way, then another. We prayed. We cried. And finally we decided—on to Atlanta!

I think I was the one who really decided. I remembered my picture in the newspaper with my back to the camera, and my promise to myself that I wouldn't stop until I could show my face and look right into the camera.

It was the spring of 1968. I remember I was wearing a blue sweater and skirt. Arlie Apple, Mrs. Apple's husband, drove us in the Apples' car. Mr. and Mrs. Apple sat in the front seat. I sat in the back seat with Mama and Daddy. It never crossed my mind that it was unusual for a teacher to take time off from her work to drive one of her pupils in her own car to see a plastic surgeon in another state. I just accepted it at the time as perfectly natural. I didn't know enough then to recognize all the effort and sacrifice that other people were making to help me. I recognize it now, and I don't think I'll ever be able to express all my gratitude.

I was terribly excited as we drove out of Chattanooga, and at the same time scared to death. It was my first trip away from home. We stopped at a restaurant on the way, and I had my first meal in a restaurant. I was fascinated by the menu—I'd never seen one before. I'd never given much thought to restaurants and how they worked. Daddy had sometimes stopped at a hamburger stand during our drives and brought a hamburger and Coke to the car. But I'd never been *inside* a restaurant, or seen the long list of foods and prices.

"What's à la mode?" I asked. "What's a porterhouse? What's a cocktail?" I insisted on reading the menu aloud, asking about the foods that weren't familiar to me.

"Hey, look at that—chicken two seventy-five! Is that a whole chicken?"

"No, baby," Daddy explained, "that's just one portion."

"But that's more than Mama pays for a whole chicken," I protested. "This place is a gyp!"

The grown-ups all laughed at my excitement over the menu, and the laughter made us all a little less nervous. Of course, we ate at an off time and went to a booth way in the back, where I wouldn't be noticed. In Atlanta we stopped first at the house of relatives. My father had a brother there, Coy Fox. Coy and his wife, Juanita, have four children, three boys and a girl. I visited with my cousins, but all during the visit I was nervous and worried about what lay ahead.

As we got closer to the doctor's office, we were all so jittery we hardly knew where we were. On Peachtree Street, Mr. Apple hollered at a truck driver for directions. "You go to this block, the next block, then left and left again," he started to tell us. When he saw how puzzled we looked, he said, "Just pull in after me and I'll show you."

The surgeon had an office in a tall building. His waiting room was bigger and more crowded than Dr. Barnwell's, and I had a hard time enduring the stares and whispers as we waited. I got a book and buried my head in it. When the doctor, a compassionate, kindly man, looked at me, he shook his head a little sadly and said to Daddy, "I'm sorry, Mr. Fox, but I'm afraid I can't take your daughter as a patient. You see, I'm nearly ready to retire, and it wouldn't be fair to her because she's going to need long-term care."

When he saw the disappointment in our faces, he said

quickly, "But I have an associate, a very competent man, and I'm going to have him look at her right now." His associate, Dr. William E. Schaffer, examined me and asked me a lot of questions and made an appointment for me for surgery about two months away.

It was July, very hot and humid, when we drove to Atlanta again. The doctor worked on my nose to improve the airway on the right side, and he removed a bulge at the nostril. (I have to call it a nostril for lack of a better word. But it was really just a big opening.)

At the hospital I was in a room with several younger children and had no choice but to be brave and set a good example for them. How could a grown-up twelve-year-old make a baby of herself in front of little kids who hadn't even started school yet? I remembered how I'd cried for Mama and Daddy to stay with me in the hospital when I was little like them. Now I made myself be brave and played nurse to the small children, telling them, "Don't be afraid, your mommies are coming back soon."

I didn't tell the little kids, "You know I really hate the green suits that the doctors and nurses wear in the operating room." It wasn't right to tell them that, but to this day I hate green, because it's the color of the operating gowns and the caps the doctors and nurses wrap around their heads and the paper covers they put over their shoes. I don't like to wear green, because every time I see it I'm reminded of the dreaded operating room. I didn't tell the little kids about that. I just closed my eyes when I saw a surgeon or an intern go by in green. Usually I was sedated or partly out by the time I got to the O.R., so I didn't have a chance to think about running away at the last minute.

When I got back to Chattanooga, Dr. Phillips did something I'd wanted very much for a long time, but he'd been putting off until my mouth and gums had finished most of their growing. He pulled out those two ugly, stubby top front teeth that jutted straight forward and made me a little bridge of three pearly white teeth that went up and down like other people's teeth.

Now, for the first time in my life, I could smile. In fact, I found myself smiling all the time—not just because I wanted to, but because I couldn't help it. My upper lip, pieced together to correct the clefts I was born with, was not long enough from the red part up to where the nostrils should have been to cover my new teeth. The following November, on my next trip to Atlanta, the surgeon lengthened my upper lip a little so it would extend better over my front teeth.

At last, I could smile, a real smile of sorts. You can be sure I did plenty of smiling. When Mrs. Apple said, "Debbie, you're so pretty when you smile," I became the smilingest girl around. I liked to see people's faces light up when I smiled at them. I hadn't realized before what a handicap it was not to be able to smile. Once I could smile, I felt more cheerful in just about every way, and the people around me—Mama, Daddy, my sister and brothers—seemed a lot more cheerful too.

But soon there was nothing to smile about. Just before we were to return to Atlanta for more surgery, my doctor there, the younger one who had replaced the doctor who was retiring, had a heart attack and died. Suddenly—just like that—a strong, healthy-looking man in his middle years! My second doctor to die! We weren't as attached to this doctor as we had been to Dr. Barnwell. Even so, it was a harsh blow—as if somebody were trying to put obstacles in my way, to make what was already hard even harder.

The doctor who died had a young assistant who came right out and said to us, "I can't do any more for Debbie than has already been done. To reconstruct her eyes and nose requires really advanced surgery, and only a few surgeons in the whole world have that kind of skill."

He went on to explain that he'd seen an exhibit at a medical meeting of the work of a Dr. Edgerton in Baltimore and he thought I should see him. He wrote out his name for us—Dr. Milton T. Edgerton—and told Mama she could get in touch with him at Johns Hopkins Hospital in Baltimore, where he was chief of plastic surgery.

Once more we went back to Chattanooga and thought real hard. Should we start now with another doctor, our fifth after two had died, one had retired, and one had thrown up his hands? More travel, more strangers, maybe more heartbreak. Were we meant to try so hard?

Mama and Daddy were back and forth with indecision. They were both in very low spirits at this time. Atlanta had been too much for them—all the effort of going there and working with new doctors and getting our hopes up— and then have it all come to nothing! They were beaten down in spirit. Daddy looked so sad and tired. I know he thought the whole world was against us.

I wasn't discouraged—not yet. I was sure of what I wanted. I wanted to be fixed up so I could go to school and have my picture taken for the yearbook with Karen and Kathy and the other children in my class. Mrs. Apple backed me up. She was the one who wrote to Johns Hopkins and found out everything she could about Dr. Edgerton. "He's a very important surgeon," she told us. "He's known all over the world. I think he's the right one for Debbie. I have a feeling that he's the one."

As I look back now, I can see that Mrs. Apple was like a mother cat moving her kittens around by the scruff of their necks. She knew we *had* to go to Baltimore to see this doctor. I can still see the determination on her face and the flashing of her brown eyes as we went round and round the problem. Yes, she was the mama cat, bent on carrying her kittens to safety.

Of course, we went to Baltimore. If we hadn't, I'd still be shut away in our house, faceless and hopeless. I wouldn't be writing this book. I wouldn't have a book to write. I wouldn't have a face—or a life.

# ∾ 8 ∾

# Promise in Baltimore

On January 27, 1969, Mama, Daddy, Mrs. Apple, Mr. Fitch, and I headed out of Chattanooga in a snowstorm and drove the many miles across the Appalachian Mountains and through Virginia to Baltimore. I was put in a room on the third floor of the children's wing at Johns Hopkins Hospital. Mama and Daddy were given a beautiful suite at the John F. Kennedy Institute, a special center for the diagnosis, evaluation, treatment, and rehabilitation of handicapped children. The Institute is right near Johns Hopkins, and the lovely guest suite is used by important visitors who come from all over to study the work being done with children. Sometimes it's also used by parents of patients.

We thought we'd be in Baltimore for only two days, maybe three at the most. But our stay stretched out, as one test followed another and doctor after doctor took his turn studying me. At the end of three days, Daddy had to go back to Tennessee or be in danger of losing his job.

Mama checked out of the fancy suite and moved into a motel a few blocks from the hospital. One of her problems was finding a laundromat and keeping us in clean clothes, because we'd just brought enough for a couple of days.

I was full of a great sense of anticipation about meeting Dr. Edgerton. I knew he was very famous and very important. Even before I ever saw him I had a feeling he was going to mean a lot to me. He walked into my room the first time with a nice warm smile on his face. He tilted my face up toward him with his hand. I turned my head a little to the left so I could see him with my good right eye. As he continued to look at me, with that warm, friendly smile on his face, I felt a little rocket of happiness go up inside me. I don't know how to describe the feeling except to say that all at once I knew I was home safe. This man would not let any harm befall me. More than that, he was going to take me out of the shadows and lead me into the sunshine. I knew all this about him in that first, almost wordless exchange of glances.

Dr. Edgerton is a tall, very handsome man with fine features and what I would call a patrician look. That's a word I've never used before, but I've read it in books and I think it describes the great dignity and look of leadership he has. He was born in Atlanta—you know that right away from his softly Southern voice. All I can say is that his very presence is reassuring. Busy as he is, with an operating schedule almost completely filled three years in advance, he always has time for a few personal words with a patient. He has a way of letting you know that you're important to him as a person, not just as an unusual case.

Much later I saw Dr. Edgerton's curriculum vitae, the list of his degrees, academic appointments, and honors. It covered two typed pages, and from it I learned that he was born in 1921. His father was a doctor. He went to medical school at Johns Hopkins, was a captain in the army medical corps in World War II, and worked at Valley Forge General Hospital, where doctors repaired the broken faces and bodies of wounded soldiers. He was a resident at Johns Hopkins in surgery, then in plastic surgery, and he went on from instructor in plastic surgery to professor and head of the department of plastic surgery at Johns Hopkins. He was a member of twenty-four different medical societies, and had served on more than a dozen important national committees and editorial boards—things like the American College of Surgeons, Graduate Education Committee, and the American Board of Plastic Surgery. He's been a visiting professor of plastic surgery at the Christian Medical College in India; the University of Rochester School of Medicine; Washington University School of Medicine; the University of Chicago School of Medicine; Yale University School of Medicine; and the University of Toronto.

Well, if I was looking for a new experience, I certainly found it in Baltimore. On Friday, the first day the doctors worked on me, I had what they call an EEG, a test that measured my brain waves. I was given anesthesia for the test, and I had a terrible headache afterward. Saturday and Sunday there were no tests, and I had a lot of time on my hands.

There were many children in that part of the hospital, some of them my age. I thought at first it would be fun to talk to them and play with them, but it didn't work out that way. Those children gave me a real rough time. They made faces at me and rolled their eyes outward to mock the way my eyes went. One of the boys kept shouting at me every time I stepped into the hall, "Watch out, here comes the monster!"

I was glad Mama and Daddy weren't around when those kids called me names, because I think it would have broken their hearts. I wanted to turn my face to the wall and cry and cry. But I decided that no matter how much I hurt inside, I wasn't going to let them see it. So I held up my head and ignored them. There was a lovely woman there named Anne Blaustein, who was the coordinator of the Jaw and Facial Clinic, which Dr. Edgerton headed, and she told those heartless children a thing or two when she caught them taunting me. That slowed them down, but it didn't stop them.

One way I passed the time on that long, long weekend was to roam the endless underground tunnels that connected the many spread-out buildings of Johns Hopkins. An attendant or a nurse always went with me, and they showed me how I could walk six blocks in one direction and four blocks in another. I pretended to myself that I was exploring a cave. I imagined that when I turned a corner I would discover the treasure room. I'm sorry to say I never found any treasure.

What I did find was a pay telephone, and one day I decided I would call Joyce. I was feeling lonesome for her. The fact that I'd never used a pay telephone before didn't bother me a bit. I borrowed a dime from a nurse, put it in, dialed the operator, said I wanted to make a collect call, and gave her Joyce's number. It was the easiest thing in the world to do. For some reason everyone seemed to think I'd done something remarkable.

Joyce kept saying, "Debbie, you mean you placed this call all by yourself? Mama didn't help you?"

"Mama isn't here," I told her. We had a nice visit on the phone, and I got over my attack of homesickness. But to this day Mama and Mrs. Apple talk about that first time I used a pay phone. I think maybe they were worried that I'd get the habit and start running up everyone's phone bill.

On Monday the tests started all over again. An electrocardiogram that day, eye tests and dental examinations the next, endless probing, poking, injecting, photographing, measuring, X-raying, questioning. In his Jaw and Facial Clinic Dr. Edgerton had brought together a plastic surgeon, a neurosurgeon, hearing and speech experts, a dental surgeon, an eye surgeon, a psychiatrist, a geneticist, and various other experts. All these highly trained specialists had their turn with me.

I think you'll begin to understand why I fell in love with Dr. Edgerton at first sight when I tell you that the very first thing he wrote in that long official medical record was: "Debbie Fox has a delightful personality with much personal charm and she writes excellent poetry." I don't tell you this to pat myself on the back—but I do want to share my wonder that a renowned specialist can be so very human in his reaction to one young afflicted child.

I wish I could describe each of the doctors to you and tell you what each one did and said. But that crowded week of my life at Johns Hopkins is just a blur to me of doctors' faces, most of them kind and concerned, of nurses and social workers who did their best to make me comfortable, and of cruel, tormenting children—all of them with defects or serious medical problems of their own, or else why would they be in the hospital? But there they were, those unhappy children turning their fears and angers on me because I looked worse than they did.

Years later I got to look at some of my medical records and I found out what those important doctors said about me. The speech specialist found that in spite of all the torn-up tissue I had no speech, language, or hearing disabilities. At times my voice was quite nasal, and that troubled me a lot. But there was no question that I could talk and make myself understood, and my hearing was perfect. Back when I was very small, a doctor had once told Mama that, given the shape of things in my nasal passages and in the back of my mouth, it was going to be impossible for me to speak. Poor Mama—that was something else for her to worry about. But, luckily, nobody told *me* the bad news. So I just began to talk—starting very young—and I've been talking away ever since.

I've been singing, too. I have a good musical ear and can sing anything I've heard. I can also pick out on a piano or small organ the melody of any song I've heard.

The dental surgeon reported that I had remarkable occlusion, everything considered, and I was doing well with the partial bridge of front teeth that Dr. Phillips had made for me.

The eye surgeon noted that the cornea on my left eye was scarred—probably from the time when that lid didn't close properly. He found that I had 20/20 vision in my right eye but only something like 20/300 in the left, and that while I could use my left eye, I depended almost entirely on my right eye. That was true enough, for I'd gotten into the habit of turning my head in such a way as to focus my good right eye on anything I was reading or writing. But I never had to wear glasses, either then or later.

The psychological evaluation found that I was tense and depressed at the prospect of more surgery. I can't argue about that. I'd already had thirty-six operations and was looking ahead to I didn't know how many more.

The social worker asked me how I felt about my face, and I looked her straight in the eye and said, "I'm tired of it."

The genetic specialist took all kinds of tests, counted my chromosomes, and asked Mama at least a million questions about herself and the family as far back as she and Daddy could remember. In the end, he reported that I had no specific genetic disorder. The best he could figure out was that something had gone wrong inside Mama's uterus while I was developing. Some cells had failed to divide or had divided the wrong way and a pinpoint error in the developing fetus had multiplied and multiplied to cause the massive deformity in my face and other parts of my body. Why had it happened? He had no explanation—some error, some injury, some insult—he just didn't know.

He noted on the record that Mama was thirty-seven and Daddy forty when I was born. Older parents sometimes have children with Down's syndrome, which results in the defect called mongolism. If you've been wondering all along whether that might be my problem, I have to assure you that it never was. I did not have any of the physical symptoms of a mongoloid child. And I certainly had none of the mental retardation. I suppose I should point out that very often children with severe facial defects like mine are taken for retarded or brain-damaged. Sometimes they're put away at birth into institutions for the retarded. (I met one such little girl I'll tell you about later.) I was told at Johns Hopkins that my IQ was 120. That's certainly not retarded.

At one point Dr. Edgerton asked me, "Debbie, what bothers you most about the way you look?"

"It's my nose," I told him. "I hate that messed-up nose."

"All right. And what's the next most important thing?"

"I guess I'd have to say my eyes. I want my eyes fixed."

"And what else?"

"I'd like my eyebrows made straight and put where they belong."

"And after that?"

"My upper lip—it's all tight and funny."

"And then?"

"The way I talk. It should be clearer."

"Anything else?"

I brought up from the fold of my dress, where I'd kept

it hidden, my right forearm, which ended in just a stub. "This hand needs fixing real bad."

It's funny that I answered that way, because when I was very small my hand had bothered me most of all. I used to hold it out in front of me and say, "Hand, get fixed. I need you." But as I got older and understood how bad my face was, the hand slipped to the bottom of the list.

Maybe this is the time to give you a better explanation of what was wrong with my face. First, there was the condition known as ocular hypertelorism. This occurs once in every 180,000 births, and it means that the eyes are too far apart when you measure from the center of one pupil to the center of the other. It's not just the eyes themselves that are wrong, it's the whole skull; there is too much bone in the center of the face and the eyes get pushed off to the outer part of the forehead.

I also had a bilateral cleft palate that extended through my upper lip and my hard and soft palates. The clefting had torn me up so much that the center of my face at birth was hardly more than a hole. What nose I had lacked proper cartilage and inner structure.

I didn't really understand about all these defects and how they were connected until a friendly nurse on one of my hospital visits let me see some pictures in a book about embryology—the study of the baby before it's born. I found myself staring at a strange, tadpolelike creature with a high, rounded forehead, a slit for a mouth, and two widely separated small black circles that looked like a frog's eyes. The nurse said the circles would later become nostrils.

"Where are the eyes?" I asked.

"They're way off on the side of the head, where you'd expect to find the ears," she said. "See these two tiny swellings? They're not ears—they're going to form eyes."

What I was looking at was an embryo at five weeks. At six weeks the drawing showed the nostrils already turned into little pits, but still far apart. At seven weeks I could see the beginning of a wide, pushed-in nose, with deep furrows from the nostrils down to the corners of the mouth opening. At this stage the eyes were still just oval swellings and still widely spaced. But they'd slid far enough forward so that I could see them on the front of the face.

The embryo didn't really begin to look like a human baby until the tenth week. Then the eyes and nostrils were still wide apart and there wasn't a nose yet, but the furrows in the upper lip had disappeared. "Every human being goes through these stages of development before birth," the nurse explained to me.

As soon as I saw those drawings, I knew that whatever went wrong with me had happened by the tenth week of fetal development. It made me think of a scene on a television screen when the camera suddenly stops and a moving picture turns into a still picture. For some strange reason, the development of my face must have been interrupted around the tenth week of my mother's pregnancy. As a result, my eyes didn't come together as they should. My nose didn't take shape. The halves of my palate never came together. Those heavy furrows in my upper lip, instead of smoothing over, tore apart. And the tear reached far back into the roof of my mouth and into my palate.

"Now look at these drawings of the arms and legs of a six-week-old embryo," the nurse said to me. I saw a hand that was just a circle with bits of cartilage inside.

"The cartilage you see there," she went on, "will soon form the palm and the fingers. Then at eight weeks the tiny fingers will be fully formed—like a miniature of a future hand."

I could see right away that in my case something had gone

wrong in those same early weeks, so my right hand never got to be a hand.

What happened? I guess that's going to be the big unanswered question as long as I live.

When all the specialists finished examining me, I sat up on the stage of a lecture room at Johns Hopkins Hospital and more than a hundred doctors, nurses, and technicians reviewed my case. They asked me questions. They asked each other questions. They argued back and forth with each other. I felt like a prize dog at a dog show, with the jurors debating out loud whether the tail was furry enough or the legs long enough.

I'll tell you a secret, though. While in some ways it was an awful ordeal, something to be endured and gotten over with, in other ways I liked it a lot. I liked being the center of so much attention. For such a long, long time I'd been cowering in the back of elevators, ducking into the last booth in restaurants, wrapping my face in a scarf, peeking out of the back window of the car, hiding, hiding, hiding.

Now, here I was up on the stage, almost like a star. There were all those doctors, some of them specialists honored all over the world, putting all their attention on me for just one reason—to give me a new face.

I felt warm and cared for inside, the way I did when Daddy took me on his lap. I didn't feel the least bit shy or frightened. I was actually enjoying myself.

I remember that one of the younger doctors came up to examine me at close range. He was really handsome, and I couldn't stop myself from saying in a loud whisper, "Hey, he's a good-looking one." The young doctor blushed. All those other doctors, even the most serious ones, laughed. A few even applauded.

"Oh, Debbie? She's everybody's friend," Daddy said to Anne Blaustein when she told him how I'd made those big-shot doctors laugh.

But that was the last laughing we did for quite a while.

# 〜9〜

# *"We Want to Move Your Eyes"*

Dr. Edgerton called our little family into his office to explain the seriousness of what we were getting into. Daddy had come back from Tennessee to take us home, and we all sat there, tense, worried, but full of hope, too. We all knew that this was it. My fate was in the hands of these great doctors. If they could help me, maybe I really would have a face. Maybe I really would be able to lead a normal life. If they couldn't help me—well, then we'd just have to accept the fact that God wanted it that way.

I felt tense and at the same time peaceful, if you can understand that. I won't say I wasn't worried. But under the worry was a feeling of safety and coming home. The words of the Psalm went through my head: "He leadeth me beside still waters." I was ready for whatever I would hear from Dr. Edgerton, good or bad.

"What we're planning to do, Debbie," he said, "is to start with your eyes."

He saw that I was about to say something, but he raised

his hand to calm me and went on. "I know you want your nose worked on first. But the eyes are basic. Until we move your eyes closer together, we won't have a structure to work on. It's not just your eyes, but the whole bony socket around your eyes—all that must be repositioned closer to the center of your face. In a way, we're talking about re-shaping your skull. Until that's done, we're not going to have any success in building you a nose or working with the other soft tissues."

Daddy had reached over to take my hand. I could feel his hand tighten on mine now. "I'm not sure what you mean about reshaping her skull," he said.

"One of the main problems that Debbie has," Dr. Edgerton explained in his soft, gentle voice, "is that there is too much bone in the center of her forehead. That's why her eyes are so far out of position. What we must do is cut away some of that bone, quite a bit of it, and then move the eye orbits—that's the eyes and the bony rings around them—closer together."

I could hear Mama gasp and catch her breath. "Is that dangerous, Doctor?" she managed to ask.

Dr. Edgerton smiled at me, a kind of half smile, and nodded his head. Then he turned to Mama: "I'd have to say there's danger, Mrs. Fox. There are serious risks. There are risks to the eyes because we will be making incisions close to vital muscles and nerves, close to the optic nerve itself. There's a possibility—I think it's remote, but it's still a possibility—that she might lose her sight. That's one of the things I have to tell you."

Mama went white. Daddy's hand clamped on mine so hard that I almost had to cry out.

"Beside the still waters," I repeated to myself.

"There are other dangers," Dr. Edgerton went on. "In cases of this type there is reason to suspect that the brain may not be in the proper position. It may lie too low—down behind the eyes instead of up behind the forehead. In that event, we'll have to raise it to where it belongs. As you know, there are risks whenever we do surgery inside the skull. It's not likely—but there could be damage to the brain or there could be infection. Brain damage—I don't have to spell out what that means. Infection could mean meningitis, even death."

That last word—death—just hung in the air after Dr. Edgerton spoke it. We were very quiet for quite a long time. He gave us time to take in what he had told us.

Mama finally had a question: "Doctor, what's happened in other operations like this?"

Dr. Edgerton took a long time answering. Then he said, "We're talking about a kind of cranio-facial surgery that's still in the pioneering stage. Until very recently we had no answers at all for children with problems like Debbie's. They had to live with their deformities. But I've operated in the last year or so on two children with conditions like Debbie's. Their problems weren't as severe by any means, but they were similar. And I've done comparable surgery inside the skulls of several adults. The results have been very good. Very good indeed. I think there's an extremely good chance that my team can give Debbie a reasonably normal face. I wouldn't attempt surgery of this kind unless I thought it would be successful. But the decision is up to you."

Mama and Daddy looked at each other. Then they looked at me. Then they looked back at each other. They didn't seem to be able to talk. The words wouldn't come. Dr. Edgerton turned to me:

"What do you think, Debbie? What do you want?"

"I want my face fixed," I told him, quietly but firmly. "I've always wanted to be fixed up."

I had no trouble making up my mind. It wasn't even a hard decision for me. I believed God wanted me to be put right. I believed Dr. Edgerton was the one who could do it. Beyond that, I wanted so badly to look normal and to lead a normal life that I didn't care about the risks.

I suppose risks are easy to face for a child, who doesn't really understand them. My own death seemed so unreal to me that I didn't even think about the possibility. Blindness, infection, brain damage—they were unreal too. What could a thirteen-year-old know about such things? So I was all for plunging ahead.

For Mama and Daddy it was much harder. Tragedy is real for grown-ups, I realize now. They'd lived through enough not to be misled by their own optimism—particularly after all they'd been through with me. Miracles don't happen all that often, so they had the good sense to be cautious.

There was also the problem of the cost of all this surgery—surely it woud run to thousands and thousands of dollars. When Mama asked Dr. Edgerton about that, he said not to worry. "There are not going to be any bills from the surgical team," he told us. "What we're going to do for Debbie will open opportunities for hundreds of other children. We wouldn't think of charging you."

Mama and Daddy heaved a great sigh of relief after that. Dr. Barnwell had almost never sent any bills, either. The doctors in Atlanta had been paid out of Mama and Daddy's savings. Blue Cross, which we had from Daddy's job, took care of my hospital room. So now the only expense would be the travel to Baltimore and meals and hotels for Mama and Daddy.

We made an appointment for surgery in June so we could go ahead if we wanted to. Cranio-facial surgery is extremely complicated and requires the coordinated work of many specialists who have complicated schedules of their own. That's why the big operations are always scheduled long in advance—weeks, months, sometimes even years.

Then we went back to Soddy and began to worry.

# ∾ 10 ∾

# *Time for Hard Decisions*

I returned to school in my own living room. By this time I was in junior high and had three telephones, one to my home room, one to my English room, and another to my math room. I had taken my workbooks with me to Baltimore and worked on them between all the tests. But I was a little behind and had to study extra hard to catch up. Luckily, I was so busy that it saved me from some of the agonizing the grown-ups around me were going through.

It was probably the worst time in Mama and Daddy's life since the first awful weeks after I was born. Mama got pale and lost a lot of weight. Daddy was nervous and jumpy. My room is way off in the back of the house behind the kitchen. Even so, I could hear them talking deep into the night, and many mornings Mama's eyes were ringed and red. Sometimes Mama got up in the middle of the night and sat at the dining-room table writing long letters to Madge Apple. I think she tried to sort out her thoughts by putting her fears and terrors on paper for Mrs. Apple.

Mama was afraid I might die. She was afraid I might go

blind. She was afraid I might end up even more disfigured. She was afraid my brain might be damaged and I would turn into a vegetable. None of her fears was neurotic or unreal. There was a perfectly good chance that any of these desperate things could happen, as the doctor had explained. After all the effort, all the suffering, I might end up worse off than before.

"Are we playing God?" Mama wrote once to Mrs. Apple. "Are we taking His work into our hands?"

Mrs. Apple had her own moments of doubt. I could feel her looking at me when she was teaching me history or going over the long division that was my worst enemy in school. Her eyes bored right through me as if she were trying to get to some secret, some hidden knowledge buried deep inside me.

"You're not afraid, Debbie, are you?" she'd ask. I knew she wanted me to be strong and resolute. I knew that if I hesitated—if I hesitated for even a minute—the whole plan would fall through. If I gave up, then everybody would give up. We'd send our thanks and our apologies to Dr. Edgerton. The operation would be canceled. And I would go through life with my ruined face, covering it, twisting it sideways, looking down at my feet.

I don't think anybody alive ever knew her feet as well as I did. I'd gotten into the habit of looking down when I was in a public place and couldn't turn my back. That way, people wouldn't have to look at me, suck in their breath, and turn away. My toes were among my best friends. They would continue to be all the rest of my life if we decided not to go to Baltimore.

Mama wrote one of her letters to Mrs. Apple on March 12. Next to the date she put the time, 4:15 A.M. She wrote, "I am trusting you will pray for us to keep a sane mind as we decide what to do about Debbie. Her friends and her contacts with the outside world are so limited. She hardly sees anyone and gets to go out very little. Her life and mine are somewhat like hermits'. I don't want her to slip into an imaginary world. I know she's worrying about the trip. Pray for us to hold out."

Mama was right—in spite of all the progress I'd made, we were still living almost like hermits. Yes, I had my schoolwork and my telephone hookup. I could talk with the children at lunchtime. Kathy and Karen came by to see me once in a while. Or my friend Darla dropped by after school. Sometimes I went to Joyce's or to one of my brothers' houses—but not all that often.

There were family parties at Christmas, usually at our house. We didn't bother much with birthday parties in our family. My big treat was still that drive with Daddy every afternoon when he got home from work. My only other outings were to the doctor's or the dentist's—and you couldn't call those visits very entertaining.

It wasn't much of a life—that was for sure. Mama was as locked in as I was. She didn't drive at that time, and anyway, Daddy took the car to go to work. So there we were, shut in together with plenty of time to brood. Now I realize that Mama must have done a lot of brooding about the future. At thirteen, I wasn't very mindful of the future. Mama surely was, and I now know how much she worried about what would become of me if anything happened to Daddy or her. Who would care for me after Daddy and she were gone?

I didn't even have church to comfort me. Mama still wouldn't let me go. She went to church every week, but I don't know how much comfort she got out of it. She begged Daddy to start going to church with her. He held back. He wasn't a churchgoer. Something kept him from going. Mama had to go alone.

Daddy never put his fears about me into words, but I could tell by the way he looked at me, by the way he seemed to pull me toward him with his eyes, how deeply disturbed he was. He could lose me—lose his baby. He surely didn't want that. The way to make certain he wouldn't was to call off the operation. But if he did that, I would be robbed of any chance at a whole, normal life.

What a terrible decision to have to make! The agony continued all through that spring—a particularly beautiful one in southern Tennessee. I wrote some more poems that spring. I wrote a wistful one about the tiny, whirring hummingbird.

### A HUMMINGBIRD

A hummingbird is a tiny thing
But such a joy it will bring,
Just to sit and watch it nip
The flower's nectar and sip.

I sit each day and watch it
When it swoops down and sits
Upon the flower to gather food.
It seems to be in such a happy mood.

It's such a pretty, colored bird
It can't be described by any word.
I like to watch how fast it toils
As though the food would surely spoil.

It comes in early spring.
Many joys it can bring
To fill sad hearts and minds
When friends seem hard to find.

There was a squirrel scampering around our yard, and I wrote a poem about its lively antics.

## A Squirrel

Did you ever watch a squirrel,
As he sits on his nest in a curl,
Away in a tree so high
Where nobody can come nigh?

Sometimes he comes down to the ground
When nothing is around
And carries food to the tree.
He acts as if on a spree.

With nuts in his little jaws
Sitting and looking at nuts in his paws
As if he doesn't know what to do
With these other two.

He packs them away for winter
No matter if he's a beginner.
By nature he knows how it's done
But still he's just begun.

He knows with winter months ahead
He can rest in bed.
When there's plenty of snow
He can lie still and grow.

I was touched by Mrs. Apple's extraordinary devotion to us. She had taken my problems and made them hers. And I wrote a poem to her.

The dearest thing on earth to me
Is dear friends to go and see
When in a troubled time
You call on them—everything is fine.

I learned this in early years
When in schooltime my fears

Were great and I would cry.
Mrs. Apple would comfort me and say

"Debbie, listen and don't be sad
Look at me and really be glad
You have someone to stand by—
Look and listen, time will fly."

Mrs. Apple taught me six years
I learned to love school through tears,
And she was dear to me
This love from me will never flee.

But I wasn't entirely wrapped up in my own worries. This was the spring before the first landing on the moon, and the newspapers and the news broadcasts were full of stories of the astronauts and their preparations for the great feat that was soon to be one giant step for mankind. I wrote a poem entitled "The Planets."

### THE PLANETS

When we look at the moon and stars
And all planets including Mars
We wonder and say—maybe
God meant for us to try and see
What wonderful material He made
And the colors seem to be jade.
We watch the stars twinkle at night
And the moon shines so bright
We know there are other places
Because scientists have found traces
Of them all when they took
Their telescopes and looked.
We know there are seven more.
As we have said before
They are great studies for man

As we study and try to plan
A way to find out everything
About those planets that our earth rings.

I did a lot of praying of my own that spring. Every night, on my knees before I got into bed, I prayed to God to strengthen me and to comfort Mama and Daddy and to guide the hands of the doctors who were going to fix my face.

I never wavered—I knew I would be fixed up. I'd been promised. I believed. "He leadeth me beside still waters."

# ∽ 11 ∽

## Green Is for Operating Rooms

I entered the hospital on Monday, June 9, 1969. This time we knew we'd be in Baltimore for a while and took enough clothes. Friends and neighbors had given me presents, and there were a lot of mysterious boxes. We opened some of them as soon as I got to my room. I was delighted with the gifts of scarves, nightgowns, fluffy slippers, handkerchiefs, and other nice things. There was one box I was especially curious about. It looked like a hatbox, but Mama wouldn't let me open it. She said it was a surprise for later.

The night before the operation Mama washed my hair with a special preparation given her by the nurse. She dried it with a towel, and I saw that she wasn't rolling it up the way she usually did to give it a wave and soft curls at the ends.

I guess I should tell you that my beautiful blond hair was my pride and joy. It was long and silky, and I loved it when Mama shampooed it for me. If your body is ugly and your face is misshapen and only your hair is beautiful, then all your longing for loveliness centers on the one lovely

thing you have. My hair was precious to me above everything else.

"Mama, you forgot to roll up my hair," I told her. I
didn't tell her that I wanted it to look particularly nice
for Dr. Edgerton in the morning.

I suppose I should have noticed that Mama looked a
little uncomfortable. But she just kept on drying, and said,
"You don't need it put up tonight. We'll do it after the
operation."

I let it go at that because I had something else to occupy
my attention. My roommate was a little black girl who was
in the hospital for some kind of surgery—I never did find
out exactly what. In Soddy and in the farm country around
it there are very few black people. I had seen blacks on the
street in Chattanooga and in Atlanta when I went to the
doctors' offices. But I had never seen a black person close
up. I'd certainly never been in the same room with one.
Now here I was with a little girl in the next bed and her
worried parents hovering over her.

If Mama and Daddy were taken aback by such a close
introduction to integration, they didn't let on. I wouldn't
be telling the truth if I said that it came to me in a flash
in that hospital room that there is a connection between
what some people have suffered because of color and what
others have suffered because of deformity. I wasn't having
big thoughts like that. But the idea must have lodged somewhere in my unconscious mind. Much later, when I
learned more about the problems of race and color, this
picture always came to me: two little girls in hospital
beds, peeking at each other around a partially drawn curtain to see just how the other one was different.

After my hair was dry, Mama and Daddy left for a few
minutes to get some supper. I was too restless to stay in bed,
so I put on my robe and slippers and went out into the
hall to have a look around. The children's wing was

brightly decorated with posters and cutouts on the walls. As I looked into the rooms I could see lots of dolls and stuffed toys on the dressers and windowsills, waiting to be played with. Some of the children were asleep. Some were sitting up with their arms or legs in pulleys. A few waved at me as I went by. I waved back.

The children on this corridor were either younger or sicker than the ones on my earlier visit to Johns Hopkins. There were no mean boys making faces at me this time, no cruel shouts of "monster." I was glad of that, for in the hospital in Chattanooga and in the one in Atlanta I'd had no trouble with the other children and I'd really been more comfortable with strangers in the hospital than anywhere else. We all had something wrong with us there—we were more equal—so I usually strolled around without worrying what people might think or say.

As I walked to the end of the corridor that night before my operation, I heard crying. I looked into a room, and there was a little girl of about five who was sobbing as if her heart would break. Then the seven-year-old in the same room chimed in.

"Hey, cut out that noise," I said to the smaller one. "You're too big for all that blubbering." I picked up a rag doll from the dresser and put it in her arms. "And you, too." I said to the other one. "You're *much* too big. What's this all about, anyway?"

They stopped sniffling and began to tell me that they were both to have their tonsils out in the morning and they were scared. Just then a nurse looked in, but when she saw I had quieted the children, she backed away.

"Tonsils aren't anything," I told the little kids. "You're in and out in no time. And you get to eat a lot of ice cream."

I'd just about gotten the little girls settled down when Mama appeared in the doorway with a look on her face like I'd never seen before. "Oh, there you are, Debbie, thank God, thank God!" she said, and rushed into the room and hugged me to her. Daddy came in right behind her, and there were some nurses behind him, and they were all grabbing at me and saying, "Thank God, thank God."

I didn't know what they were carrying on about until Mama told me she and Daddy had come back from supper and found me gone. Because *they* were so afraid for what lay ahead for me the next day, they thought *I* was afraid too. They thought I'd run away. They pictured me running through the unfamiliar streets of Baltimore . . . lost . . . running . . . lost. Mama and Daddy were still shaking as they led me back to my room.

I tell you, sometimes parents just don't have enough faith in their own children.

The operating room had been reserved for me from seven in the morning until midnight. I was given an injection in my room and don't remember being taken out of it. The last thing I said to the nurse was "Please have a Coke ready for me when I get back here."

The operation was easy for me because I was unconscious. But it sure wasn't easy for the others.

The day was hot, sunny, and endless. Mama and Daddy were staying again at the Kennedy Rehabilitation Institute. Very early in the morning, they arrived at the family lounge in the old Johns Hopkins building where the operating suites were. Ed Fitch, the assistant superintendent of schools, who had driven us to Baltimore, was with them. Madge Apple had planned to fly in after the operation, but the day before, she'd talked to Mama and Daddy on the phone, and I guess she found them so shook up that she was afraid they weren't going to make it through the night. So she caught the first plane she could. She arrived at the Baltimore airport at two A.M., got to the hotel at three, took a bath, and waited for daylight. She was too keyed up even to try to sleep. She joined the others in the hospital waiting room in the morning—without having been to bed.

Well, as I've heard it, they paced and they fidgeted. They adjusted the air conditioning and readjusted it. Mama prayed quietly. Daddy, the one who didn't go to church, prayed out loud. Mr. Fitch, white as a sheet, murmured over and over, "Don't let her die. Don't let her die."

They all took turns going out for coffee and bringing each other sandwiches. Mama says she counted the flowers on the drapes a million times. They picked up magazines and put them down—no one could read. A preacher from some kind of Indian church came in and prayed with them. A Church of God preacher came and they prayed again. Other people who had relatives in surgery came and then left as the operations on their loved ones were completed. Morning turned into afternoon, afternoon to evening, evening to night. Still they waited. Occasionally Mrs. Blaustein stopped by to tell them that the operation was still under way. The night was hotter than the day. And the suspense mounted.

Inside the operating room, I was the only one who wasn't aware of the drama that was going on—because I was unconscious. The others were tense and keyed up far beyond usual O.R. excitement. This was no ordinary operation. Never before had surgeons intervened so boldly to reshape a badly formed human skull. And never before had surgeons dared such a drastic rearrangement of a human face as to move a child's eyes more than two inches closer together. There was nothing in the medical literature to guide them.

(I probably should tell you at this point that, unknown to Dr. Edgerton and the others at the time of my operation, in Paris Dr. Paul Tessier, a distinguished plastic surgeon, was performing surgery similar to what was done on me. But his work had not then been reported in this country. Later Dr. Edgerton and Dr. Tessier met, and Dr. Edgerton sponsored Dr. Tessier's membership in the American College of Surgeons.)

In fact, the procedures were so new and untried that Dr. Edgerton and Dr. George Udvarhelyi, the chief neurosurgeon at Johns Hopkins, had tried to anticipate some of the problems that might arise by dissecting the skulls of several cadavers. The practice runs did not help a great deal, because those were normal skulls and my skull was abnormal. In spite of all the X-rays and the tests and the measurements they'd done on me in advance, there was really no way for the surgeons to know what they'd find when the operation began.

About three dozen people came and went in that operating room on that extraordinary day. Dr. Edgerton was the only one who was there almost continually from beginning to end. His residents in plastic surgery took turns with him during the day. In the morning, he worked with Dr. Udvarhelyi, the neurosurgeon, and his assistants. Later in the day, he worked with Dr. David Knox, the senior eye surgeon, and his assistants. The operating room nurses came on and off duty in shifts. The anethesiologists and their assistants came and went in shifts. The blood people worked in shifts. A medical artist was in and out of the operating room, sketching the high points. And a movie cameraman worked to record the key stages of the surgery.

Everyone who came in at any point had to scrub and put on green coveralls from head to feet.

Just about everything that took place in the operating room that day presented special problems. The problems began with the anesthesia. Everyone knew this would be a very long, drawn-out procedure. To keep a patient anesthetized for eight, ten, maybe twelve hours is an extremely tricky business. One of the problems is that the anesthetic gases are very dry—dry as desert air. If the lungs take in this dry air over a long period, the patient can be in trouble. There can be lung complications and possibly pneumonia.

So, in my case, the anesthetists bubbled moisture through the gases all during the operation to prevent damage to my lungs. They warmed the gases with an electric coil to bring them up to my body temperature. Since this humidifying and warming were not usual procedures, they required special monitoring all during the surgery.


Working with Dr. Udvarhelyi, Dr. Edgerton began the operation by opening my forehead. The plan of the operation was to remove about two inches of bone from the center of the lower part of my forehead. This was the excess bone that caused my eyes to be displaced too far to the right and the left. Once the center bone was out, the plan was to free the bone around the eye openings, so that the bony orbits, with the eyes inside them, could be swung closer to the centerline.

The cutting of the bone began in that middle area between my eyes. Dr. Edgerton used a tool called an osteotome, which is a surgical chisel that he taps with a lead mallet. It is somewhat difficult to use a drill or a power tool in this area, because any buildup of heat could be damaging to the brain.

To make this effort extra complicated, the skull in that area is not a single thickness of bone. The frontal sinuses are located there; they are a pair of cavities with a back bony wall that is located right up against the brain. This back wall cannot be clearly seen on X-rays. The earlier practice sessions had been devoted to finding out the best way to cut through the back wall of the frontal sinuses. The doctors were trying to be extra careful because they were working so perilously close to the brain. Not only was there the risk of damage to the brain itself; there was also the possibility of leakage of spinal fluid if the covering of the brain was damaged. And if the brain covering was cut, there was immediate danger of infection and meningitis.

As the doctors had suspected, my brain was lying lower than it should have. The frontal lobes were located behind that extra bony space between my eyes instead of being positioned up behind the forehead.

One of the complications that Dr. Edgerton feared most did happen as he and the neurosurgeon cut through the back bone of the frontal sinuses. Some of the dura, the smooth covering of the brain, was adhering to the back bone and a bit of the dura was torn as the bone was cut away. There was a small leakage of spinal fluid.

A tremor of alarm went throurgh the operating room. Dr. Udvarhelyi quickly cut some fibrous tissue from the area around the temple and used it to patch the dura. The leakage stopped and the brain was no longer exposed. Everyone breathed a deep sigh of relief.

The next step was to free the bones around the eyes so the eye orbits could be moved centerward. At this stage, Dr. David Knox, the eye surgeon, moved into position next to Dr. Edgerton. The big problem here was to avoid cutting into the brain as the osteotome chipped away at the bony brow. Dr. Edgerton solved this problem by slipping his forefinger into the opening made by the removal of the center forehead bone. He positioned his finger between the brow bone and the brain. His finger inside the cranial cavity could then guide the osteotome and protect the brain from the sharp tip of the instrument.

Dr. Knox kept track of all the muscles and ligaments around the eyes to make sure no harm came to them. And he watched the various retractors that were holding tissues in position to be certain there was no damage to the eyeballs or to the optic nerve. Finally, the last bit of bone around the eye area was cracked apart. Some of those final fragments of bone were so close to the eyeballs that those in the operating room held their breath. At last the bony orbits of the eyes were free. The operation was at its crucial point.

Could the eye orbit be swung far enough centerward to get my eyes into a normal position without kinking or pulling on the optic nerve? This was the question that had no answer. Until now.

Dr. Edgerton and Dr. Knox carefully moved each eye orbit toward the center, working first on the left side, then on the right. They wired the bones together at the center of the forehead with steel wires. Before that, they cut away a bit of the bone at the bottom on each side at the center to leave room for the nasal passages.

My tear ducts had been dissected out earlier. Now they were put back in position and allowed to run through the new bone area. All sorts of complicated things had to be done to anchor my eyelids in their new positions and keep them operating properly. This included making new ligaments from fibrous tissue from the nose area. Bone grafts were placed at the outer sides of the eye orbits to keep them in their centered position. My eyeballs and optic nerves now occupied the space where the frontal lobes of my brain had been. My brain was finally safely lodged where it belonged, up behind my new brow.

As the long, long day stretched into evening, Dr. Edgerton took the surplus skin that had covered the bony area in the center of my face (the bony part that had been cut away early in the morning), shaped this skin and tissue into a kind of cone, and stitched it in place to form the beginnings of a nose for me.

I lost a lot of blood during the operation. That doesn't really say it. I lost *all* my blood during the operation. It was continually replaced during the surgery, first with whole blood, later in the day with fresh blood, because fresh blood contains a vital clotting ingredient. Without that clotting ingredient, I might have bled to death during the surgery.

By the time the surgeons were finished with me, they had taken hundreds of stitches in my brain covering, in the tissues both under and around my eyes, in my forehead, and on my newly constructed nose. They had reduced the distance between the centers of my pupils from 103 to 65 millimeters—a good two and one quarter inches.

When they finally wheeled me into the recovery room, everyone in the O.R. was exhausted to the dropping point. Still in his green operating suit, well splattered with my blood, Dr. Edgerton went down to the waiting room to tell Mama and Daddy and the others that all had gone well.

It was nearly 10:30 P.M. by then.

Mama and Daddy were practically trembling with terror. They'd never dreamed the operation would take all that long. They were sure that something terrible had happened and nobody had wanted to tell them. Even Mrs. Blaustein's repeated assurances had not calmed them.

Now Dr. Edgerton came and dropped with weariness into a chair. "Debbie's fine," he said in his soft voice. "She's going to be all right. Everything went beautifully."

All Mama could do was fall back in her chair and say, "Thank God, thank God."

Daddy insisted on shaking Dr. Edgerton's hand. "Thank you, Doctor, thank you for saving her," he repeated over and over.

Mama and Daddy wanted to see me right away. But Dr. Edgerton saw they were at the breaking point and ordered them off to rest immediately. Then he let Mrs. Apple and Mr. Fitch peek into the recovery room for just one minute.

I guess I must have looked like a mummy. My head was all wrapped in bandages. There were dressings in my mouth. Various tubes hung from me. I heard a nurse say, "You can't talk, Debbie, but if you recognize anyone's voice, nod your head."

I heard Mr. Fitch say, "Debbie, you're beautiful."

I nodded my bandaged head.

Then I heard Mrs. Apple say, "I'm here too."

I nodded again.

Then they were gone. and I sank into a deep, deep sleep.

# ～12～

## *It Was a Miracle!*

Mama came in later and spent most of the night sitting in a chair next to my bed. She prayed, she slept, she worried. "I never knew a night could be so long," she said later.

Sometimes she lookd at me, all wrapped up in those bandages, with tubes going in and out of me, lying very still and quiet, and she became afraid I was dying. It would be her fault for letting them go ahead with the operation. Then she imagined me blind—with my face still ruined, and blind on top of it—groping my way around the house. That would be her fault too.

"Dear God," she prayed, "please don't let that happen. This child has suffered enough."

I knew it was morning when I heard the nurses talking and moving around in the room. But I couldn't see them. I couldn't see Mama, either. Everything was black in front of me. I began tearing at my bandages in terror with my good hand. I was never so frightened in my life. I couldn't see. I knew I was bandaged, but that didn't really register in my mind. I guess I was still foggy from the heavy anesthesia. What did register was that I couldn't see.

Dr. Edgerton had said that if the optic nerve was stretched too much or torn, there was a chance I might lose my sight. I was frantic that this had happened, and kept pulling wildly at my bandages. I had already tugged the outer wrappings off before Mama could grab my hand and stop me. A nurse ran for a doctor. He came in, put back the bandages, and assured everyone that I hadn't done any real harm to myself. Then he laid down the law to me:

"If you touch those bandages just once more, Debbie, I'll have to tie your hands down. Do you understand that?"

I nodded, but pointed frantically to my eyes to show him I couldn't see.

"It's the bandage. You know you can't see through a bandage."

The nurse who was working with him tried to be helpful. "As soon as the bandage is off, you'll see as well as ever," she said. "Maybe even better."

I wanted to believe them. But deep down I knew they were as worried as I was. They had no way of really knowing. And nothing to go by. Never before had eyes been moved this much—a full two and a quarter inches closer. It was such a big move and so many things could go wrong. The doctors had told me that the operation had worked earlier on two children whose eyes weren't nearly as far apart as mine. Maybe it was good only for smaller adjustments—to bring eyes about an inch closer. Two and a quarter inches! Maybe that was too much dislocation for the eyes to take.

I thought of how much misery a single cinder can cause, and here I'd had everything in the eye region moved around. What was more, maybe, as some people had feared, it was too risky working so close to the brain. Maybe the surgeon's finger inside my skull had accidentally done some small but irreversible damage. Dozens of things could have gone wrong during the long hours I was on the operating table. So there was the chance that my vision could be gone altogether. Or it could be distorted. Or seriously impaired.

Now there was nothing to do but wait and pray. Wait and pray.

I prayed quietly, "Please, God, make my eyes all right. Please, God, make the work of my doctors a success."

I was frightened. But I couldn't believe God would let me go blind. Mama and I had sometimes sung hymns together, and now the words of one of them came back to me:

> I will not let go until You bless me,
> 'Til You bless me,
> I'll hold on, and on and on;
> Here's my cup, Lord, fill it up;
> If it takes from now til midnight
> Or 'til early morning's hour,
> I'll keep holding on, dear Jesus,
> Until I feel Your power,
> I will not let go until You bless me,
> 'Til You bless me,
> I'll hold on and on,
> I'll hold on, I'll hold on.

So I held on while the bandages stayed on my face for several days. I mostly drifted in and out of sleep during that time. The suspense continued. I was alive. My mind was working. But could I see? Did I look better? Were my eyes now in the right place? And what about my nose? I tried not to worry. Then, suddenly, I had something else to worry about.

Mama and Daddy were sorting out the piles of gifts that had come for me after the operation. I heard Mama ask, "Ed, is that the box with the crocheted shawl over there?"

"No," Daddy said, "that's the one with the wig."

I heard that word "wig" and I just about went crazy. I remembered, then, the hatbox that Mama wouldn't let me look at. And I remembered how she hadn't rolled up my hair the night before the operation. That was it! No question about it.

My hand flew up to my head. I couldn't feel through the bandages whether my hair was still there. But I knew it was gone. It had been cut off! I let out a screech that must have frightened the whole floor of the hospital.

My hair, my beautiful hair, was gone! I could feel the tears sliding down under the bandages. Now my spirits really hit bottom. Without my hair, I had nothing. It was the one thing about myself that I liked. "Mama, they cut off my hair! How could you let them?" I wailed. Mama tried to comfort me. Daddy petted me. I was inconsolable. Nobody could calm me down.

The whole wig business, I found out later, had started weeks before my operation when Joyce said to Mama one day, "Do you think they're going to shave off Debbie's hair?"

In one of those middle-of-the-night letters she was sending to Mrs. Apple, Mama wrote, "Do you reckon they will shave any or part of her hair off when she's operated on? If you think so, we'll have to get her a wig. I wouldn't want to bring her back on the bus like that. I didn't want to ask about shaving her hair when I was talking to you on the phone, I was afraid she'd hear me. But when you call me, I can answer in a way she won't know what I'm talking about."

Bless Mama—do you know what else she wrote in that letter? She asked, "Did your lilac bush bloom this year?

All mine were so pretty." Just think—right in the middle of all her worries, she could take time off to notice a lilac bush. I guess I'm really a lot like Mama.

Anyway, Joyce and my brothers bought me a beautiful wig, and it was waiting for me in that hatbox I wasn't supposed to know about until Daddy made a slip and spilled the beans. Maybe he did a good thing after all, because I made such a nuisance of myself over the lost hair that nobody had much time to worry about my eyes and all my other serious problems.

Finally the day came when Dr. Edgerton removed the bandages. He was especially cheerful and friendly that day. "Well, Debbie," he said. "Let's have a look at those pretty eyes." He unwrapped the bandages and as the last of them came off, I blinked desperately as the brightness of the room almost overwhelmed me. Actually the room, a small examining room on the same floor as my hospital room, was in semidarkness. But after your eyes have been covered over for a while, any light seems bright. Now I was dazzled by the green walls, the white curtain by my bed, the faces of Dr. Edgerton and Dr. Knox and the nurses, who were all looking at me so hard you'd have thought their lives depended on it.

"Hey, what's everyone staring at?" I asked.

"That's our Debbie," Dr. Edgerton said. "We're staring at you."

The doctors looked at my eyes very closely. They made my eyes follow the beam of a small flashlight. They held little cards with letters in front of me and made me read the letters forward and backward. They moved and shifted various objects in front of me. As my eyes followed the objects, as both eyes focused on the objects simultaneously, the doctors became absolutely jubilant.

I had my sight! It was a miracle. None of the things we'd dreaded so much had happened. My vision was as good as ever—no, it was better, for now I could see straight forward and I could see with both eyes.

I wanted to give everybody in sight bear hugs, but Dr. Edgerton calmed me down and told me, "You'll have to stay quiet for a while until the healing's finished. No jumping around, young lady!"

Within the next few hours I think the entire medical staff of Johns Hopkins came by to have a look. Everyone smiled a lot and congratulated me, and things were just fine until I asked for a mirror. Nobody wanted to give me a mirror. But when I put my mind to wanting something, I usually get it.

"I want to see a mirror right now," I insisted.

Finally they gave in and let me have a mirror. The nurse warned me, "Debbie, you're all swollen and bruised. You've had the most major surgery that's ever been done on a little girl's face. So you're not looking one hundred percent yet."

"I'll settle for fifty percent," I told her, and took the mirror in my hand. I held it off to the right, as I usually did. But instead of seeing the right side of my face, all I saw was the edge of the mirror. Slowly I moved the mirror toward my left. As I moved it, I could see more and more of my face coming into the glass.

When the mirror was directly in front of me, for the first time in my life I saw my whole face all at once. It was all stitched over and puffy and bluish in spots. That didn't bother me at all. The amazing thing was that I could see my face, my whole face, by looking straight ahead.

I brought the mirror closer to me. Then I extended it out at arm's length. When I did that, I began to feel almost a little dizzy as my two eyes focused together for the first time and adjusted to close-up and more distant vision.

I was very excited at that, because I knew right away what it meant. Now I'd no longer have to hold a book off to the right so I could read it with my good right eye. Now, for the first time, I'd be able to hold a book straight in front of me, the way everybody else did.

It was a joyful moment for me as I realized I'd taken another step on the road to normalcy. I let myself anticipate the pleasure of holding a book or a picture in front of me—of looking straight at the TV set with both eyes—of looking at Mama and Daddy with both eyes.

I could feel the happiness well up in me.

I also saw something else astonishing as I stared in the mirror. My awful flap of a nose had been replaced by what looked to me like a big false nose. It was much too big for my face. I didn't care. It was a new nose, made out of the extra skin from my forehead, and it was in the place where a nose should be.

Then I caught sight of my head—bald except for a little fuzz. Mama was all ready for more crying and hysterics from me. I surprised her. I'd gotten the whole thing out of my system by then. All I said was, "Hey, that's some skinhead I've got."

Mama then opened the hatbox and gave me the wig that Joyce and my brothers had bought for me. I tried it on and looked at myself in the mirror. It was a beautiful wig, with lovely silky hair just the color of mine. But it was too long and full, and I knew I'd have to have it cut when I got home.

Everybody was surprised at how little pain I felt after the operation. With pieces of my skull cut out, my eyes moved around, and a new nose sewn on, you'd think I'd be screaming in agony, but I wasn't. I'm not going to say it didn't hurt at all—because it did. My head ached pretty badly at times, and my face pulled and itched. But it wasn't unbearable. It never hurt enough for me to cry about the pain.

I once read years later that corrective plastic surgery seems to hurt less than other kinds of surgery because the patient has a positive, optimistic feeling about the results. I'd like to believe that's true—but I'm not so sure about it, because later, when I had to have my right hand strapped to my stomach for three weeks to make a skin flap for a new finger, and again when I had a piece of tissue cut away from my thigh and sewn into my face to raise the cheekbone area, I had really terrible pain—far worse than the big operation.

One of the last nights I was in the hospital, I had trouble sleeping. I was allowed to be up and around by then, so I went out into the hall to see what the nurses were up to. There was a very young nurse who had become my friend, and she asked me, "Debbie, would you like to see what the operating room is like? I can take you there to have a look."

I've told you how much I hated the operating room and the green outfits that the doctors and nurses wear there. But I'd never really seen the place, because I'd always been put out before I'd been wheeled in. Now I liked the idea of actually seeing the room where Dr. Edgerton had performed the big operation.

The nurse took me upstairs in the elevator and into a small room where we both slipped into those green operating gowns. The nurse explained that we had to wear them to avoid contaminating the room. I wrapped the gown around me. She tied it in back for me and put a surgical mask on my face. We tiptoed in as if an operation were going on. The room was dark and spooky until she touched a switch and a huge overhead light went on that flooded the room with so much brightness that I had to blink for quite a while to get used to it. There was a table under the light, which I recognized as the operating table. Next to it was another table with instruments lined up and ready. We tiptoed across the tile floor. I looked around at the tile walls and white ceiling. We didn't dare touch anything; if we did, it would have to be sterilized again.

When I got close to the operating table, I closed my eyes so I could imagine myself lying there with all the important doctors and their assistants standing over me, waiting their turn to work on me. I thought of the six main doctors, their assistants, several anesthetists, and I don't know how many nurses. They had been on their feet, all of them, from early morning until nearly ten at night. They had not stopped to eat.

The atmosphere must have been unbelievably grim and tense as they opened my skull, exposed my brain, relocated it to a more correct position, broke apart the bones of my forehead like so many children's building blocks, took some of the bone away, rearranged other parts of it, moved my eyes within their bony orbits toward each other, re-adjusted the delicate muscles around the eyes, wired the bones back into proper position, took the skin on the center of my forehead that was left over when the excess bone was removed and reshaped it to form a nose, sewed it all together with those unbelievably fine stitches the plastic surgeons use, bandaged me, and then waited, during days of suspense, to see whether or not all this heroic effort had been worthwhile.

Of course, that night when I visited the operating room in my trailing green cotton gown, I didn't know all the details of my surgery. I've learned most of them since. At that time, all I knew was that a never-before thing had been done and that forever after my life was going to be different—and better.

A shiver went through me as I left the operating room with the nurse, the kind of shiver you get when you're in the presence of something truly awesome. "Thank you, God," I whispered, "thank you for giving the knowledge and strength to my doctors to fix my face."

## ∾13∾

## *The World Discovers Me*

The big thing that happened at this point in my life was not just my recovery, but something totally unexpected. The world discovered *me*.

The chief of the Associated Press Bureau in Baltimore, a man named Marvin Beard, sent out a news story about my operation. He wrote:

TEEN-AGE GIRL'S FACE REBUILT IN 13-HOUR OPERATION Baltimore, June 13 (AP)—Doctors at Johns Hopkins Hospital, using some techniques for the first time, have taken a major stride toward rebuilding the face of a 13-year-old Tennessee girl in a 13-hour operation.

Six surgeons, working from 8 A.M. until 9:20 P.M., Wednesday, were involved in the case, which Dr. Milton Edgerton, a plastic surgeon and one of the six, called "only the second of its kind I have ever seen in 20 years of practice."

"Some of the techniques we used had never been

done before, at least not in this country," Dr. Edgerton said.

It was the 37th—and by far the most major—operation for the girl, Deborah Fox of near Chattanooga,
who, although she has never attended classes a day in
her life, will be in the eighth grade this fall.

Deborah is an amateur poet, has an I.Q. of about
120, and wants to work, as a career, with handicapped
children. She is the fourth—and youngest—child of a
Chattanooga foundry worker.

The operation literally changed the configuration
of Deborah's skullbones, moving her eyes from the
sides to the center of her face, and began rebuilding
her nose. Brain surgery, eye surgery and plastic surgery were involved.

Still to come are operations on her nose and mouth.

When Deborah was born, she had virtually no face.

"I would almost say the condition was unique,"
Dr. Edgerton said. "The variations were different
from any I had seen previously. There were no books
to read about such an operation."

The article went on to tell how Mrs. Apple taught me
at home and how I used my special telephone instead of
going to school.

The story ran in several hundred newspapers across the
country. You wouldn't think those few words would change
my life so completely, but they certainly did. The *News-
Free Press* in Chattanooga printed the AP story and then
ran an editorial which said: "This little girl has shown a
greatness that would be exemplary to the rest of us who
had not had the mountains to climb that she has faced
and still climbs. The teachers and doctors and friends who
have helped her with love and skill and in so many other
ways also set an example that is most laudable."

Me? Debbie? They had to be talking about somebody else!

The next Sunday several ministers in Chattanooga preached sermons about me. Dr. Harry L. Mercer of the Northside Presbyterian Church told his congregation, "Such a wonderful deed as the history-making operation performed at Johns Hopkins on a thirteen-year-old girl named Debbie Fox did not issue from hearts of malice. This miracle was wrought by God through hearts of compassion and concern, through minds dedicated to medicine."

I was still in the hospital when the calls began pouring in, from newspapers, from radio stations, from TV commentators, asking for more details about what had happened to me. Mrs. Apple offered to take care of the phone calls, letters, and telegrams. I think she could have used two secretaries, she was that busy. Of course she had to say no to people who wanted photographs of me or personal appearances. I was in no shape to be interviewed, but I did read the mail. I couldn't believe the letters I was getting. Debbie Fox of Soddy, Tennessee, the girl who was born with such a deformed face that she'd had to hide in her own house for thirteen years, was becoming a kind of celebrity. People were writing to me the way they write to movie stars. I had to pinch myself to believe it.

My bed was piled high with letters. One from Massachusetts read, "I thought I had my health problems, but what you went through must have been hell. You see, I have asthma, but it's getting better. It's awful not being able to breathe. I really admire you for all your courage. I think you'll turn out beautiful. You've opened the door to many hopeless people."

Another, from a girl studying to teach the deaf, said, "I think your story is absolutely fascinating, inspiring and awesome. I think you are a very special person carrying out such high goals."

One from a great-grandmother said, "My nine children, 25 grandchildren, nine great-grandchildren were all born with normal bodies and I want to thank God."

A mother and her daughter wrote, "Praise the Lord! and thank Him for letting us know about Debbie and His goodness to her."

A Marine in Vietnam wrote, "Don't let the doctor put one of those sticks in your mouth until you know who ate the ice cream."

Not all the letters were from strangers. One was from Mrs. Brainard Cooper, who had worked in the nursing office at Children's Hospital in Chattanooga when I was little and went there to have my face worked on. She recalled that Dr. Barnwell had once said to me, "Debbie, this is Mrs. Cooper, say hello to Mrs. Cooper."

"I shall always remember," Mrs. Cooper wrote in her letter, "your shy, sweet look when you repeated 'Hello, Mrs. Cooper.' You were so pleased you could say it. I stooped down and kissed your cheek and hugged your sturdy little body. We all loved you so very much."

I could feel my eyes getting misty as I thought about Dr. Barnwell. He had wanted so much to see me whole. If only he could have lived for this wonderful day!

Then I could hardly believe it—there was a letter from President Richard Nixon, written from the White House. This is what the President of the United States wrote to me on June 19:

Dear Deborah,

I read recently of your great courage and strength in undergoing the latest of many operations and that despite the confinement that these operations have brought you will be going into the eighth grade this fall. It is particularly heartwarming to see that you would like to work with handicapped children when you grow up. I know these must be difficult days for you, but your own high spirits and determination to succeed assure me that the day will come when you will reach your goal. Mrs. Nixon joins me in sending our best wishes to you for the years ahead.

Sincerely,

Richard Nixon

It was unbelievable. Mama and Daddy kept passing that letter back and forth between them. There was wonder on their faces. A letter from the President of the United States to their little girl! Not in their wildest dreams had they imagined such a thing. I'll never forget their love and pride at that moment. Me, I just kept pinching myself to make sure I was awake and not dreaming.

I found out later that it was Marvin Beard who was responsible for the letter from the President. Mr. Beard had a friend who worked for Vice-President Spiro Agnew. The friend suggested that the President might like to know about the amazing operation at Johns Hopkins. I guess the friend figured Mr. Agnew would be proud to have people hear about the great thing that was done in his home state of Maryland. Mr. Beard sent the President a copy of his news story. A few days later he got a brief thank-you note from H. R. Haldeman, the President's aide, and he thought that was it. He was as surprised as I was when the President wrote to me.

And I was as surprised as any of them to read in Mr. Beard's story and in the President's letter that I wanted to work with handicapped children when I grew up. I'm going to be honest with you and tell you that up to that time I hadn't given a single thought to what I wanted to do when I grew up. I wanted my face fixed up and I wanted to get married like Joyce and have babies. I'd never thought seriously about a career for myself.

A real career was beyond my dreaming, but not beyond Mrs. Apple's. Mr. Beard had asked her a lot of questions about me and my plans, and that teacher of mine, always scheming, told him, "'Debbie has so much sympathy for children who are handicapped. She has a special feeling for them after all she's been through. I wouldn't be surprised if she decided to work with handicapped children."

Mrs. Apple, as usual, was putting a bee in my bonnet. She didn't know that she'd have the President of the United States helping her at it. Since then, I *have* worked with handicapped children in summer jobs. Maybe, when I finish my surgery, that will be my career.

Meanwhile, I'm running way ahead of myself.


Dr. Edgerton and all his colleagues were delighted with the way I was healing. In fact, he caught us completely by surprise by coming into my room one morning and saying, "Debbie, you know what? I'm sending you home tomorrow. You're finished here for now, but we'll see you again in a few months."

That was fine with me—I couldn't wait to get home. But Mama had a problem. "Is it all right to take her home on the bus?" she asked the doctor. No, it was not all right—Dr. Edgerton was very firm about that. Daddy and Mr. Fitch had had to go back to Tennessee as soon as my bandages were off. Mama and Mrs. Apple were still in Baltimore with me—but they had no car. Dr. Edgerton said no jarring on a train, plane, or bus. I had to go by car, but who would drive us?

Mrs. Apple's kind husband, Arlie, came to our rescue.

He had retired from his job at TVA by then and could get away. He started out from Chattanooga at five in the morning. He'd only gotten as far as Dayton when his car came to a stop. He managed to roll it into a service station to look for help, but the station was closed. A police car drove up, and when the policeman saw him peering into the station at that hour, he thought a holdup was going on. Luckily, he only asked questions. When Mr. Apple explained the problem, the policeman helped him fix up the loose water hose that was giving him trouble. Mr. Apple finally made it to Baltimore, and drove me home very slowly and carefully, making sure I didn't get bumped or jolted.

As soon as I got home, I called my sister and said, "Come see me, I'm beautiful." She didn't need an invitation—she was already on the way. My sister, my brothers, my nephews rushed over to see the "new" Debbie. Cousins, aunts, and uncles came to see me. There were reporters and people from the school system, and now and then strangers knocked on the door. So many people came trooping through our house that I thought the porch steps would be worn out.

Nearly everyone said the same thing: "Debbie, you're beautiful."

It was great to hear. But it wasn't strictly true. I was still far from beautiful. My eyes were much closer to normal, that was for sure, and they looked forward rather than off to the side. But they still needed work, and the eyebrows were still askew. My nose was no longer that awful flap hanging up between my eyes. It was a nose now, but only a rough draft of a nose. At this stage it didn't have any cartilage or understructure, so it had a fake look. And you could see clearly where it was stitched on. The hollows and curves in my cheeks were still in the wrong places. My upper lip still had a pieced-together look. And, of course, my hand was still missing.

I wasn't a sight you'd really care to see. But compared to where I'd been—compared to what someone had once described as "that awesome hole in that little girl's face"— I was looking pretty neat.

So I accepted the admiration and the congratulations. Sometimes I said, "My doctor should take the bows. He's the one who did it." Or else I said, "You ought to congratulate Mama and Daddy. They had a worse time on this than I did." Or else I said very quietly, "I always believed God would have me fixed up."

I never saw Mama and Daddy looking so happy. Their worst worries and fears were now over. We had all gone through a terrible tunnel together. Through the valley of the shadow. Now we were out in the sunshine. Daddy hugged me to him every time we passed. Mama hummed as she went about the house.

Even the mailman looked happy when he dropped big piles of letters on our porch. Yes, the mail was still coming in from all over—people congratulating me, reaching out with their warmth and love, asking for the names of doctors who might help their afflicted children, pouring out their own pain and misery. Every day brought more mail. There were gifts, too: books, handkerchiefs, sweaters, things people had sewn or knitted for me. Quite a few of the letters contained money—sometimes a dollar bill, sometimes a check for ten or more dollars with wishes for my speedy recovery.

In September, there was a return trip to Baltimore to let the doctors evaluate their work. When they looked me

over, they gave me and themselves an A in everything.
They were just delighted. I was on my way with a face—
a new face.

"Debbie," Dr. Edgerton said to me, "do you have any
idea how many children will benefit by what you've gone
through?"

"Do you have children?" I asked him. I'd been wondering about that for some time.

"I have four," he said, "two boys and two girls. I'd like
you to meet them someday."

"I'd really like to, because I want to tell them how lucky
they are to have you for a father."

"And I'm lucky to have you for a patient, Debbie."

One of Dr. Edgerton's sons has become a doctor—he is
now working as a general surgeon at the same hospital
where his father is now head of the plastic surgery department. The other son is an architect. One of the daughters is
studying architecture, the other is a writer.

There was a big first for me on that checkup trip to
Baltimore—Mama, Daddy, and I *flew* back to Chattanooga.
I'll tell you one thing right away, flying is no picnic for me.
I was scared when I got on the plane. It was a rough, nasty
flight over the mountains. We kept dropping down into
air pockets. Just about everybody was sick, and for a while
up there I didn't think I'd ever set foot on good solid
ground again.

"Daddy, I can't take it," I whimpered, and clutched the
armrests. "Tell the pilot to land, I'm scared."

Wouldn't you know: Chattanooga was fogged in and we
had to circle and circle for another miserable half-hour
before we could come down. I couldn't help thinking that
it really would be a bad joke for me to be injured or
killed in a plane crash—after all the trouble so many
people had gone to to give me a proper face. I promised

myself then that I'd never fly again. I had to break that promise later under very sad circumstances.

Meantime, two marvelous things happened that fall. Two of my dearest wishes were granted: I went to school and I went to church. And, oh yes, there was a third great thing I'll tell you about later.

# ∽14∾

# To School—for Real

Mrs. Apple was doing some more of her scheming. She once said to me, "Debbie, the greatest thing is to dream and then make those dreams real." She dreamed for herself and for me and for others.

I knew she was very fond of me and did much, much more for me than teachers usually do for their students. It wasn't that she became a mother to me—she never stood between my mother and me. But I think now, looking back, that she saw me as a way to reach *other* children. My need was so dramatic that she would use it to persuade people to stop and think about what had to be done for all the handicapped. I was her Easter Seal child all year long.

Now she told her colleagues at the Board of Education, "You know Debbie ought to be in school. She can only go so far with her telephone. She needs the stimulation of real school." She wasn't talking just about me. She was talking about other children in the homebound program. And other children with disabilities who'd be coming along in future years.

Mrs. Apple knew how desperately I wanted to go to school. A regular class, she was afraid, would be too difficult for me because of my hand, my remaining facial problems, and my very limited experience with other children. I needed a special class. And I wasn't the only one with such a need.

At the time, I had no idea of the miles of red tape that had to be unraveled. Somehow, with her usual determination and persistence and with the cooperation of Dr. McConnell, Mrs. Apple arranged for a small class to be set up that fall of 1969 in a specially equipped ground-floor classroom at the White Oak School. It was the first class of its kind in our school system.

I'll never forget my first day of school. I began at the eighth grade, not the first grade. I was thirteen, not six. I was overjoyed and at the same time frightened to death. I got up very early that first morning and dressed very carefully in a plaid skirt and red sweater. I was still wearing my wig (the hair underneath was growing back, but was too short to look nice), and I arranged the wig and rearranged it until I got it right.

Mama and Daddy were telling me to hurry or I'd be late, although both of them knew I wouldn't dream of being two seconds late for this big occasion. I think they were more nervous and excited than I was. It must be a funny feeling for parents who've had special reasons to shield and protect a child since birth to see that child stretch her wings and start to fly.

It seemed to me I was waiting on the porch for hours, but I guess it was just minutes. Finally my school bus pulled up. I turned to wave to Mama and Daddy. They were watching me from the living room. I ran down the porch steps and skipped my way across our little walk to the street. My bus was not one of the big yellow school buses, but a mini-van painted blue with the words EASTER SEAL SOCIETY lettered on the side. It was driven by a woman, and to my surprise, I saw I was the first passenger.

I climbed aboard, walked across an open space, which I learned later was there for children in wheelchairs, and sat down on one of the seats. The driver fastened my seat belt and we were off. As we drove along the winding hill roads, I now had time to think. For the first time I could feel pinpricks of fear. Would I really like school? Would the other children like me? Would I get along with them? Maybe I'd be the one who wasn't liked. Maybe I wouldn't have any friends. I'd never had a really close friend before. How would I go about making friends now?

I can tell you, the butterflies were very busy in my stomach. I began to wonder why I'd been so eager to go to school all along. It was nearly a half-hour drive to the next stop. There Lisa got on the bus—a red-haired girl three years younger than I and in a wheelchair because of a birth defect in her spine.

I could see her sitting in her chair by the road, and wondered how she was going to get up into the bus. It didn't take long to find out, because, when the door was opened, a lift was let down. Lisa wheeled her chair onto the lift, and it brought her up to the level of the bus. She then wheeled her chair into the open space in the bus and locked the wheels, and we started off.

Lisa and I looked at each other for a while, and finally she said, "Hi, I'm Lisa."

I said, "I'm Debbie."

It took us a little time to get started on conversation, but soon I learned that she had gone to public school for a while. "What's school like?" I asked her.

"It's no fun," she said. "They give you homework, and homework is yecchy."

I told her I'd done homework all along, and she wanted to know about my telephone hookup. We were just getting acquainted when the bus stopped for Lamar, a boy of about sixteen in a wheelchair. There were eight other stops before the bus filled up, and with all the stopping, the lifting up of the wheelchairs, and the tying of belts, it took us two hours to get to school. That was to be my routine for the next five years—two hours on the bus to school and two hours home. I was always the first on and the last off, because I lived the farthest out in the country.

Our classroom had been specially set up for us with big windows facing a grassy area, lots of space for maneuvering the wheelchairs, shelves for books and play material, a piano, chairs with arms that could be moved around easily, and many bright-colored paper decorations. There was a lot about the room that was a big surprise to me. Somehow I'd thought the room would be bigger, although I was soon to learn that our room was bigger than most classrooms. I had expected rows of seats anchored to the floor instead of all the open space and movable chairs. I had thought that blackboards were really black—ours were green.

There were ten of us in the class, six in wheelchairs. The youngest was in third grade. I was the oldest, in eighth grade. Our teacher, Mrs. Geraldine Huckaby, was a motherly woman who could easily understand our problems because she had been born with clubfeet, which had responded well to surgery.

School was a lot livelier than I'd thought it would be. We moved around a great deal in the classroom. Since we were in so many different grades, we'd work separately for a while over our workbooks or reading. Then we'd all get together for sewing or singing or to learn Spanish—things like "Good morning" or simple numbers. Mrs. Huckaby seemed to be just about everywhere. Even when she wasn't looking at you, she knew exactly what you were up to. Two aides helped her in the classroom, but Mrs. Huckaby really ran things.

Lunchtime the first day of school was frightening at first. The aides and those of us who could walk pushed the children in their wheelchairs down the long corridors to the cafeteria, where our group sat at a special table. There was a hush in the big cafeteria when our little parade entered. Remember, it was the first day for all the others in the school too, the hundreds of normal children, who had never seen a sight quite like us.

There was a buzz of whispers as the children asked each other and their teachers who we were and why some of us looked so funny. One of the boys in our group was twisted with multiple sclerosis. Several wore braces. My face was still very strange. The only one who looked entirely normal was possibly the sickest among us—Janet, a very pretty girl with a brain tumor.

The buzz continued for a while. A few children left their seats to come and stare at us. Then things settled down and the other children turned back to their food or the games they were playing. After that they paid very little attention to us. And we paid little attention to them. We talked and joked and played among ourselves at lunch hour. We were happy with each other's companionship.

At the end of that first exciting day, our bus picked us up again. Now we were no longer strangers. We talked and laughed about what had happened during the day. I helped some of the children in wheelchairs position themselves on the lift as they got off. "Good-bye, Debbie, see you tomorrow," they called, and my heart sang.

"Good-bye, Debbie, see you tomorrow." Such simple words, and so precious to me. They meant normal, everyday life, the give-and-take of friendship. They were magic to my ears.

Lisa was let down at her stop and called back to me, "See, I told you about homework. I was right. Yecchy."

"Yes, you were right," I called after her. "Yecchy, really yecchy." But I didn't believe what I was saying. I loved school. I loved homework. I loved talking to the other kids. I loved being with them so much that before long I found myself in trouble.

The ten of us with our various disabilities and problems needed a lot of attention from the driver, from Mrs. Huckaby, from the aides. The grown-ups did a wonderful job, I realize now, of spreading their attention and concern around. But, having been isolated for so long, I was frantic for attention. I wanted to be noticed. I *had* to be heard from. My hand was always the first one up when a question was asked. My voice always rose above the others when we recited out loud. My singing was louder than anybody's. I crowded up to the front when we got in line. As I look back, I realize I was a royal pain, and I can understand why.

I'd never had to compete for my parents' attention with sisters and brothers—mine were too old. I'd never had to compete with classmates for my teacher's attention—she'd been all mine.

While I'd known a few children to talk to, I'd never had the chance to interact with them. I'd never been the loser in a game—or the winner. I'd never known the jealousy of having a best friend go off to whisper with someone else.

When I was about twelve, I'd had a slumber party. Karen and Kathy and Darla and some of the other little girls came over with their pajamas. We giggled and talked and ate potato chips and drank Cokes and maybe slept once in a while. Mama kept poking her head in to make sure we were all right. We had fun that night. But I never went to sleep over at any of the other girls' houses. I never really entered their lives, or they mine.

I was always an outsider. I was used to that. But I was also very much pampered and protected—maybe too much. For all these reasons, I had a really hard time getting my feet on the ground at school.

There was never enough of the attention I was so greedy for. I always wanted more. I got some by working hard and doing well at my lessons. I got some in another way—by making things up and telling wild tales. I had a crush on Lamar, a good-looking boy with dark hair. One time when I wanted him to look at me, he was busy at something else. How could I get him to look at me? A little demon inside me spoke up. I listened to the demon and raised my hand.

"Mrs. Huckaby, Mrs. Huckaby, Alan hit Janet on the head with a bottle."

People looked at me all right. In fact, there was a small commotion. Almost everyone looked at me—except Lamar. He went right on concentrating on his arithmetic workbook. I repeated my story. Alan denied it. So did Janet. I wasn't a very good storyteller, and my demon didn't give me very good advice.

It was a silly story, because Alan was in a wheelchair and Janet was not. She could easily have skipped away from him if she had any reason to. Besides, Lamar didn't find the story interesting. I was scolded and told not to try that sort of thing again.

But I tried it again and again. I kept my little demon very busy. One of my stories got me sent to the principal. Mr. Brinkley was a very understanding man, and as I look back, I suppose he must have been delighted to have a disciplinary problem among the handicapped children. It just went to prove how normal we were. But I was badly shaken up at having to go to the principal's office, and once more I promised never again.

As time passed and I got more confidence in myself and my ability to get along with the other children, I heard less from my demon. One of the last times he spoke to me, I reported that the woman driving our bus had driven recklessly. She hadn't at all, and when my story got back to her, she really gave me a talking-to.

"Debbie," she said, "I'm doing this for your own good. If you spread any more tales, I'm not just telling Mrs. Huckaby or Mr. Brinkley, I'm going all the way up to the Board of Education."

That really frightened me, and it put my demon to rest for good.

Some time later, when we were between regular drivers, we had a man driving our bus on a temporary basis. One day I told Mrs. Huckaby I'd seen the driver drinking out of a whiskey bottle while he was driving. I really had seen him do that. But Mrs. Huckaby got that angry look on her face and raised her finger right in front of my nose and said, "Debbie, I don't know what we're going to do with you."

"But it's the truth, Mrs. Huckaby, honest it is."

She still didn't believe me. Why should she? She had a grim expression on her face every time she looked at me all that day.

That Friday I saw the driver at it again. As I pushed Lamar's wheelchair to the lift, I bent down and snatched up the empty whiskey bottle and kept it hidden between the chair and me. In the classroom, I put the bottle on Mrs. Huckaby's desk. She quickly slipped it into a drawer—

I guess she didn't want to start anything in front of the other children—but I could tell by her eyes that she'd forgiven me. The next week we had a different driver.

Much later Mrs. Huckaby told me a funny thing. She was so busy that Friday she forgot to tell the principal about the bottle. Suddenly, over the weekend, she remembered it and was real worried that someone cleaning up might find it in her desk. Then *she'd* be the one who'd have to do some explaining!

Mrs. Huckaby taught me for four years, and now that so many more of the handicapped are going to school and so many more teachers will have to learn to work with them, I'd like to tell you a little about her. She was born with clubfeet, with her toes pointing backward. She could walk, but she couldn't stand still—there was no way she could balance on her feet. When she was seven she went into a hospital in New Mexico. She had to stay there *three years* while they worked on her feet!

There was no teacher or classroom in that hospital, and Mrs. Huckaby got some of the older children to show her how to read and spell. She was ten when she first went to school. They put her in the first grade, but she stayed only a week. She hurried herself along until she was with children her own age. Later she married and had two children —her son is now twenty and her daughter is eighteen. She was a teacher and recreation director at Children's Hospital in Chattanooga for eleven years and then was assistant director of a private school for the severely retarded. The year before our special class was started, her job was to ride the school bus every day, and use that time to tutor hard-of-hearing children who were in the regular Hamilton County Schools but needed some extra help.

I learned about Mrs. Huckaby's very special philosophy

one day when I was feeling sorry for myself because the other children left me out of a surprise they were planning for her—a great big Thanksgiving turkey they were cutting out and coloring. Maybe the other kids just didn't think to include me. Or maybe they were punishing me for being bossy with them. Or maybe they thought I couldn't handle the scissors and crayons because of my hand. Anyway, I was sitting off by myself, sulking.

Mrs. Huckaby, who, I'm sure, knew all about the "surprise," came over to me and said, "You know, Debbie, I learned a long time ago, when I was a little girl, that if I included myself, then others included me. Once, I remember, my friends were playing ball and they didn't ask me to play because of my feet. I went up to them and said, 'I'll keep score,' and they said, 'Okay, fine.' So whenever they played games I couldn't join in, I was the scorekeeper or the umpire. And after I included myself a few times, they got used to including me.

"Debbie, do you know that in high school I got my letter in basketball—me, with these feet?" She pointed down to the heavy shoes she still wore. "I had the loudest voice, so I became the cheerleader—and I got my letter."

Mrs. Huckaby looked at me with a lot of expectation in her eyes. I knew she wanted me to do something. I thought very hard. Then I went over to the sewing corner of the room and got out the red-and-white-checked apron I was making and trimming with red ricrac braid. Sewing, I should tell you, was my absolutely worst subject. I had the same problem with it as I had had earlier with writing— just one hand and the need to train a balky left hand when I was really right-handed.

Well, I took a small piece of that red braid over to the kids and I said, "Hey, what do you think about using this for the red comb on the turkey's head?"

Little Lisa gave me a big smile and said, "Debbie, that's

a really good idea," and there I was, in the middle of the action, where I wanted so much to be.

Another time, I was in trouble with the kids because I tried to take over during music time and wanted to put on the record player only the records I liked. I had the best ear and the best singing voice, so I thought I had the right to take charge. Later that day, during quiet time, Mrs. Huckaby came over to talk to me. She said, "You feel good when you listen to music, don't you, Debbie?"

I nodded.

"That's because you have a good ear and you're good at it. How do you feel about sewing?"

"I hate it."

"Why do you hate it?"

"Because I can't do it."

"All right. How would you like it if Janet or Jill pulled your sewing out of your hand and told you, 'That's really bad. You don't know anything'?"

I could feel myself getting angry at Janet and Jill. "I'd hate it. I'd hate them."

"Maybe they feel that way when you tell them they don't know anything much about music."

I had nothing to say to that.

"You know, Debbie," Mrs. Huckaby said, "all people, not just handicapped people, have things they do well and things they don't do well. We all have to balance it out. It's all right to feel superior about some things and inferior about others—as long as we balance it. It's all right to avoid some situations where we can't keep up. And it's all right to make the most of what we do well. But we have to remember about other people and how *they* feel. Do you think you can remember that, Debbie?"

I've tried to remember it ever since.

# ∽ 15 ∽

# *"I Get Excited About the Lord"*

School was not the only new and wonderful door that opened up in my life. As I've told you before, Mama went to church every Sunday at Sale Creek. Daddy was not a churchgoer at that time. He and I stayed home and got Sunday dinner ready for the three of us, or for more if my sister or brothers and their families were coming to our house.

Daddy was a good cook. He usually made a roast or a chicken in the oven. He'd cut up the vegetables and get the potatoes ready, and I'd cook them. I'd had plenty of time when I was home alone with Mama to watch her bake. When I was about ten, she'd showed me how to mix a cake and make a pie crust. I could steady the bowl by holding it against me with my right arm. I got to be good at stirring and mixing with my left hand. I couldn't open an egg by myself or slice things, but I had no trouble with the burners of the gas stove. I could even light the oven.

I was the table setter, and I liked arranging the plates and lining up the knives and forks to make them nice and even. The dining room was always warm, even on the coldest days, because that's where the heater is. When I was little, our heat came from a coal stove in the dining room. Later we got a gas heater that really warms up the whole house and isn't any trouble at all to keep going.

When Mama's friends or one of her sisters drove her home from church, she always pretended to be surprised by what Daddy and I served for Sunday dinner. Of course she knew—she'd done the marketing. If she especially liked the strawberry pie or the chocolate cake I'd made, she'd say, "Debbie, you've gotten to be the best baker in this house!"

I was proud that Mama liked my cooking, but very unhappy that she still wouldn't let me go to church. After my big operation I was sure she'd tell me to come along with her. But she didn't.

"Why not, Mama, why can't I go now?" I asked her.

"Debbie, I'm afraid you're just not ready. Soon—but not yet."

I couldn't get her to give me any reason beyond that. She was afraid, she told me later, that I might not be made to feel welcome. She just didn't know how to make up to me for the hurt if I was treated like an outcast at church.

One unforgettable Sunday I went to church on my own.

This is how it happened. I've mentioned that Brother Lewis Hickman lived across the way from us. He is the minister of the East Soddy Church of God, one of the many evangelical Churches of God in the Soddy area. Brother Hickman is a very handsome man with an open face and piercing eyes. At that time he still had blond hair (it's silver-white now).

He's one of those men who just seems to radiate energy. He was in the Eighth Air Force in World War II, and after the army he worked in construction on TVA. In 1949 he was saved and God called him into church service. He began with only eight members, but his church grew and grew. By the winter of 1969, when I first went, he had a beautiful brick church built almost entirely by his more than 150 members.

That first Sunday I went with his wife, Juanita Hickman. Mrs. Hickman knew how much I longed to go to church. She simply said, "Debbie, come along with me. I'll drive you there and back."

Mama had already left for her church. Daddy was off doing something. I knew this was my big chance. Mrs. Hickman drove us along a winding road, through a deeply wooded section. Then we came out near the top of a ridge, and we seemed to be very high and close to heaven. The sky was very blue against the pine trees. The red-brick church was simply built but impressive, standing there with only the sky and the pines around it.

I walked in with Mrs. Hickman. People nodded in an easy, friendly way. I felt right at home. I knew I was among friends. There was so much that was new to me and wonderful to look at that I could hardly take it all in at once. First there was the building—green walls and green carpet in the aisles, many rows of pale wood pews, handsome woodwork at the front of the church. Above the offering box the words "This Do in Remembrance of Me" were carved into the wood.

Many of the women wore crocheted shawls over their printed or brightly colored dresses. The women all looked beautiful to me with their clean, scrubbed faces—just naturally beautiful, without make-up or jewelry. The men all had a calm, peaceful look on their faces. The little children looked scrubbed and content.

I knew right away that I was welcome and at home. Everyone I passed on the way in said, "Good morning" or "I'm glad to see you." Some must have heard of me before, because they said, "Good morning, Debbie," or "So glad you came, Debbie." They just took it as the most natural thing in the world that I should join them this Sunday for prayer. Not a single person stared at me. Never once did I feel anyone's eyes slide toward me to take a quick glance and then look away.

At first I was amazed that they knew me. Then I remembered that Brother Hickman had told me that special prayers had been offered for my recovery when I was at Johns Hopkins. These good people had prayed for me, and now they were happy that I was praying with them.

Then the service began. I knew Brother Hickman had a rich, powerful voice—but I had no idea how powerful until he opened it up in prayer and preaching. When he preached, he quoted the Bible so many times that I could hardly keep up with him. I knew I'd have to do a lot of studying in Sunday school to learn all the texts he was using.

As he preached, his voice got louder and more powerful. He flung up his arms. He pounded on the lectern. Sometimes he jumped down off the pulpit and went striding up the center aisle, his voice building and building until I thought his breath would give out. But it never did.

After he'd gotten to the highest pitch and we were following him breathlessly, hanging on every word, he stopped and was quiet for a minute. Then he started up speaking again, so softly that we had to lean forward to hear him. Gradually he let the power come over him and his voice mounted up in another crescendo.

It was thrilling to hear his voice. I don't remember the exact words, but he talked about faith and sin and the devil and the way Jesus loves us. He quoted from Romans and John and Acts and the old prophets and the Psalms. Brother Hickman was not the only one filled with religious fervor. Everyone else was, too. People shouted out, "Amen!"

or "Hallelujah!" or "Praise God!" when they were moved. They raised their arms as a sign to God, sometimes one arm, sometimes both.

One man began calling out in a language I didn't know. Brother Hickman stopped in his preaching and let this man's strange words fill the church. The others listened with their heads bowed. I found out later this is called "speaking in tongues" and is considered in our church as evidence of the baptism of the Holy Ghost, as it is written in Acts 2:4: "And they were all filled with the Holy Spirit and began to speak in other tongues, as the Spirit gave them utterance."

There was a grand piano at one side of the church and an organ at the other. At one point a young woman of about twenty went up to play the piano, a man played the guitar, and a woman sang—it was a gospel song, "The Man from Galilee." The whole congregation joined in, and the singing was so loud I was afraid the glass windows would break. There were special, deeply felt prayers for the sick. One woman took a handkerchief of her sick child's up to the altar to be anointed.

In his rich voice Brother Hickman asked for divine healing. He asked the Holy Ghost to make intercession under God to bring health to the afflicted, all the afflicted. His eyes met mine, and I knew he meant me when I heard his powerful words of prayer. A shiver went through me. The congregation began to sing again. Their voices soared up to the high roof of the church. I knew the holy music and words were going through the roof and straight up to heaven. My voice was as loud as anyone's as we sang together:

> I have joy that goes beyond the circumstances
>     of this life,
> It continues to excite me, tho' the days are

filled with strife;
Like a clear and cooling river, it is bubbling
in my soul,
Giving strength to journey on for heaven's
goal.
I get excited about the Lord.
I am thrilled about the promises of His word;
I want to sing the coming King,
I get excited when I think about the Lord.

The hymn was a perfect one for me; I was so excited about the Lord I could hardly stand up. I was excited about the day and about church and about being there. Later people crowded around me and welcomed me. I felt warmed by their love. The world was a wonderful place. I was glad to be alive and to be here, in church, at last, to thank God for His many blessings.


I couldn't stop talking about church when I got home. I had to tell Mama and Daddy over and over again what it was like. They couldn't calm me down, couldn't bring me back to earth. I had dreamed about church and imagined it for so long. I think somewhere inside I'd been a little afraid that the reality would not live up to my hopes. I know that was what Mama was afraid of for me. But when she saw now how overjoyed I was and that nothing hurtful had happened to me, she was just as happy as I was about the whole thing. She made me tell her in detail exactly what had happened.

"The building is so beautiful, up there on the hill," I told her. "You should see the way it stands out against the sky. And oh, the way Brother Hickman preaches—I don't think you've heard anything like it!"

Now that I'd found everything about church was even greater than my expectations, there was only one more thing I wanted—I wanted Mama and Daddy to be in that church with me. For the next few months Mama continued to go to Sale Creek and I went to the East Soddy Church of God with Mrs. Hickman. Then Mama got curious to see what I was talking so much about. One beautiful sunny Sunday she and I went to Brother Hickman's church together. I don't know that the service was so different from hers, but right away she liked the friendliness of the people and the way they stretched out their arms to welcome both of us.

"That's a nice church, Debbie, and those are really nice people," she told me. After that she went with me nearly every Sunday. Now I had only one more goal: to get Daddy to church. It grieved me terribly that he didn't go. I felt the loss for him, that he was not saved, that he did not feel close to Jesus. And I felt the loss for us as a family, that we were not in church together.

In my eyes Daddy never did anything wrong—except for not going to church. I began working on him. And working on him. I talked about nothing else except getting him to church. Poor Daddy—I guess I was too much for him. Or rather, I guess it was the Lord who was too much. At first he made excuses and kept putting me off. I didn't stop. At last one Sunday morning he put on his best suit and said, "All right, Debbie, I'm going to have a look at that church of yours."

We drove there, the three of us. I think it was the happiest morning of my life. We walked proudly down the aisle. I could feel all the approving glances at us when Mama, Daddy, and I sat down together. If I'd thought that Daddy might be standoffish about the services, I was wrong. Right from the start, the feeling seemed to get to him. It was as if the Lord had touched him. He sang and shouted "Halle-lujah!" His voice rose in the hymns. At the end the three of us were holding hands when we sang together:

Words cannot express the joy I have today,
The blessed Holy Spirit walks with me, walks with me;
With every need supplied, I'm fully satisfied,
And soon my destination I shall see.
Now I'm walking with the King on the road that leads
        to glory.
Telling of His love in gospel song and story;
What a joy divine to feel His hand in mine!
Walking with the King of Kings.

Daddy was saved. This I had done for him, after all he had done for me. Now, at last, we had a real Christian home. We were all saved.

# ∾ 16 ∾

## Growing Up Fast

I grew up fast that fall—everybody said so. Mama said, "Debbie, you're changing from a little girl to young lady right in front of my eyes!" Well, I was nearly fourteen; it was time to grow up. My body had already begun to change from little girl to young woman. Now I wore hose instead of socks. My skirts were longer, and the clothes Mama bought me were less childish.

Daddy said, "You're getting too big to call baby." But I knew I'd never be too big to be his baby.

In January, when I went back to Johns Hopkins for my second operation, I was quite a different Debbie from the year before. I walked into the hospital like an old-timer. Everybody said hello to me by name. Mrs. Blaustein, who was always glad to see me, said, "My goodness, look at our grown-up Debbie."

This time there were none of the nasty boys around who'd tormented me on my first visit. Best of all, we weren't all scared to death the way we'd been in June. The operation coming up now was not a life-and-death matter.

The doctors wouldn't be poking around in my brain, so I wasn't going to be in any serious danger. Of course, there was the usual risk and fear of surgery, but I was accustomed to that. I'm not going to tell you I was overjoyed about going into the operating room again. I knew it was something I *had* to do—the price I had to pay for getting closer and closer to having the face I'd always wanted.

I was still wearing my blond wig when I got to the hospital. My hair was coming back in—quite a bit darker than before—but it wasn't yet long enough to wear in a way that was becoming. It made me look like a boy, and there was no way I wanted to look like a boy. When Dr. Edgerton saw me, he said, "Well now, Debbie, how are things?"

I looked him right in the eye and answered, "Blondes have more fun." I heard later that he repeated what I said all over the hospital. He got a big kick out of it.

This time there were only four doctors operating on me, and it took only four hours. Small potatoes, really, after what I'd been through. What the doctors were trying to do this time was to reconstruct the lower and upper lid region of my left eye. That lid, you'll remember, was closed at birth, and Dr. Barnwell had made an opening as soon as possible so I could use that eye and keep vision in it. It was my bad eye, as far as seeing went. The idea now was to make it look better. In the big operation both my left and right eyes and lids and everything within the bony sockets had been moved closer together; nothing had been done to improve the appearance of the lids.

This time Dr. Edgerton and Dr. David Knox, the eye specialist, built up the left lids by using some of the soft, loose tissue that formed a too-wide bridge between my eyes. The doctors added the new tissue, starting at the tiny hollow in the inner corner of the eye near the tear duct. It wasn't enough just to sew the skin there. The doctors had to drill through the sides of my nose, thread new ligaments through the holes, and snuggle in the loose folds of the eyelids. The tricky part was to draw the eyelids together in such a way as to have the lids on both sides follow the curve of my eyeballs.

Next, the four-man team started to sculpt my nose to thin it, form a rigid base for it, and have it develop a healthy blood supply.

After the operation was over and I was recovering, Dr. Edgerton told a reporter, "Debbie's doing just fine. She'll probably need about four or five more operations on her nose, mouth, and cheeks. Our long-range goal—and we feel sure we will succeed—is to make Debbie presentable and able to lead a perfectly normal life."

When I read the article in the newspaper, I was very happy about the "perfectly normal life." I wasn't so crazy about the "four or five more operations." That meant it would be at least two or three more years before I was finished.

"Mama, why does it have to take so long?" I complained.

Poor Mama—she wasn't feeling any better about the idea than I was. Maybe it was a good thing neither of us knew that "four or five operations" would stretch out into at least twenty. The time of my reconstruction would be measured not by years, but by a decade.

Maybe because I had to look ahead so much in my life—to my next operation, to the time my nose would look a little better, to getting my hand fixed—I always was able to take a long view of things. I'd even written a poem about the span of time, which I called "Life."

### LIFE

Life is such a shortened span
For woman or for man,
Hard work from child to manhood

It hardly can be understood.

When as a child we ignore
Many things that are in store
For us when we grow old
Many a misfortune we have to hold.

Life is just what we make it.
Folding our hands in our lap we sit
Thinking that good luck will overtake.
A better person, some try to fake.

We all must face the facts
Whatever fate may subtract,
And learn to live with our dreams
How large, or how small they seem.

Man's days are short and full of trouble
Grief makes them seem double.
We must pray and look above
For God has promised all His love.

Luckily, I had something to take my mind off all the operations in my future, because that third wonderful thing I mentioned before was now happening.

# ∽ 17 ∾

# *The Debbie Fox Foundation*

I told you about the floods of mail that came in after my big operation. The letters that sent good wishes or enclosed gifts were easy to answer. At first, Mrs. Apple and Mama wrote the thank-you notes. Later, when I was home, I answered many letters myself. Quite a few were from soldiers in Vietnam or Korea. Some American soldiers stationed in Turkey sent me a card so big that it opened all the way across my room and was signed with hundreds of names. The soldiers wrote to tell me how lonely they were and how they longed to get home. I never really understood why they wrote to me, except that maybe they had a lot of time and they'd seen my name in a newspaper or magazine and thought I'd be sympathetic.

I did understand the letters from men in prison. I received a lot of them; I guess they wanted to reach out to me because I had been set free and they had not. I answered all the letters from prisoners. One, I remember, was there for breaking into a house, another for stealing a car. I wrote them short notes that said things like "It's good to hear from you and I'm sorry you're in prison and I hope you're out soon." Or I said, "It's good to hear from you and the money you've sent is very well received. You know you didn't have to send it to me and I realize maybe you could have used it for yourself or for someone else."

One of the boys who'd written me while he was in prison wrote later from his home. He said, "I guess you'll be glad to hear that I have something good to tell you. My wife had a baby while I was in prison and now I'm so happy because I have someone to call my own. You can be sure I'll never do anything that will take me away from my family again."

One Sunday right after church a car drove up to our house. We weren't expecting anyone, so I went out to the porch to look. The car had Pennsylvania plates. A nice-looking young man got out and came up the walk. "You're Debbie Fox, aren't you? I'm one of your pen pals. I'm Bob," and he gave a last name that I couldn't pronounce. I invited him in to meet Mama and Daddy. He'd been on vacation in Florida and was driving back to his home in Pennsylvania. "I decided I'd go through Tennessee on my way home to thank you," he told me.

"Thank me for what?"

He looked a little disappointed that I didn't remember what I'd written him. I guess it was hard for him to understand that I'd had hundreds of letters. He'd had only one.

"I was in prison for stealing a car," he said. "I wrote you and never expected to hear from you. But you wrote back and you told me that I was in there for a purpose. That you didn't know what that purpose was. But while I was there I should stop to think about what I'd been doing. Well, your letter made me cry and I took it to heart. Now I want you to know how much it meant to me."

I think about Bob quite often, even though I never heard from him again, because he reminds me of how easy it is to touch other people and change their lives. Sometimes it takes just a letter or a few words.

Mrs. Apple must have been having the same thoughts as she answered the sad letters that came in from parents of children with facial deformities of all kinds. Some had conditions called Cruzon syndrome or Apert's syndrome. These are birth defects that involve eyes spaced wide apart the way mine were, odd flat noses, and abnormal jaws. Others had tumors of the jaw or cheeks. Others had misplaced eyes or ears or misshapen mouths or double noses.

I saw only a few of those letters. Mama and Mrs. Apple didn't want to upset me with the sadness of other people. Remember, I was just thirteen at that time. They thought I'd had enough sadness of my own.

Mrs. Apple was very concerned about those letters. A great many were from poor people, without money to look for help for their child. Some were from people who had searched all over and not found an answer. Some had never even heard of plastic surgeons; it was news to them that such doctors existed. Some had been to plastic surgeons in their home towns or nearby cities and been told to take their child home and keep him happy—there was nothing that could be done.

As a teacher of the handicapped, Mrs. Apple had spent many years working with children with problems. It really distressed her to discover that there were dozens, maybe hundreds, maybe even thousands of children with facial deformities who were not getting help. The March of Dimes, she knew, worked with children with cleft palates. Those with the less common defects or with the unexplained ones like mine had nowhere to turn. There were no directories of services, no sources of information, nothing.

Mrs. Apple told me later that she answered some of those letters with tears streaming down her face. Mothers told how their children had been shunned or locked away. They told about bungled operations. About others that had failed without bungling. About hopes created and destroyed. About hours and days spent in doctors' waiting rooms with crying babies. About money used up, debts piled up. Whole families' lives were ruined. Photographs of afflicted children that would break your heart came with many of the letters.

The message was nearly always the same: We've heard about Debbie Fox. Thank God for the miracle that has given her a face. Now, do you think there is hope for my child? If so, where can I turn?

Mrs. Apple put some of the letter writers directly in touch with Dr. Edgerton, and in that first year he saw nearly a dozen children whose mothers had written me, and began treatment for them. In other cases Mrs. Apple sent the letters to Dr. Donald Russell, a leading plastic surgeon in Chattanooga, and had him suggest a plastic surgeon near where the child lived.

It wasn't enough just to answer all those pathetic letters with doctors' names. Many of the families couldn't afford to travel to a hospital or medical center. They couldn't pay the medical expenses or didn't know how to find a private or public agency to help them with funds. Many had other problems, too. I guess they got so caught up in their sorrows over a deformed child that they just kind of lost control of their lives. They told in their letters about how their houses burned down, their cars got into accidents, their insurance ran out when they needed it most. Their troubles always seemed to come in doubles and triples.

Mrs. Apple answered the letters and wept for these

people—and began to have an idea. Why not a special organization to help children with severe facial deformities? You know Mrs. Apple by now, so you know that no sooner did she have the idea than she began to take action. She made phone calls to influential people in Chattanooga. She talked to Mr. Fitch and Dr. McConnell at the Board of Education. There were meetings, memos, plans, consultations with attorneys, discussions about taxes—all the complicated and mysterious things that go into setting up an organization, especially a tax-exempt public foundation.

I didn't know anything about these time-consuming preliminaries. All I know is that one day Mrs. Apple came to the house and said, "Guess what, Debbie! We're setting up an organization to help other children with facial deformities. And we're going to call it the Debbie Fox Foundation."

It was a good thing I was sitting down. If I'd been standing up, I might have fallen down. "You're calling it what?" I asked.

"We're calling it the Debbie Fox Foundation, and it's going to work to encourage research and provide medical services to disfigured children," she said.

I could hardly believe it. A foundation to help other children. And with my name on it! I was overwhelmed. I kind of gasped for air to catch my breath. Mama and Daddy came in to hear the news from Mrs. Apple, and they were overwhelmed too. Daddy came and put his arm around me and said, "Well now, baby, isn't that just something!" and I could hear the pride in his voice and see it in his eyes.

Mama took my hand and held it in both of hers and said, "You know, Debbie, we always said there was a reason why you were made the way you were. Maybe now we're coming close to that reason."

That very same thought was taking shape in my mind. My thoughts went back to my early years, when an angry, frustrated little girl asked, "Mama, why did God make me like this, why, why?"

Mama always answered, "He has His reasons, and some-day we will know them."

Was this His reason? That I should lead other little children out of their valley of darkness? I felt the tears starting up. I didn't even try to hold them back. I didn't have to, because Mama was crying already, and Daddy was wiping at his eyes with the back of his hand, and Mrs. Apple was dabbing at her eyes with a lacy handkerchief.

We were all crying together—this time from joy and pride and not, for a change, from misery and fear.


This happened about the time we were all going to church together, so I could hardly wait through the next few days for Sunday. By the time Sunday came, people around us already knew about the Foundation because an article had appeared about it in the Chattanooga *News-Free Press*.

As I've told you, people are always friendly at our church. That Sunday I could feel a *special* warmth. In my Sunday school class, the lesson was Confidence During Crisis. We talked about the way faith in God will bring us safely through the storms of life.

I felt almost shivery as I realized that the Bible texts of the day were ment for me in a way that seemed beyond coincidence. The next was: "Be of good cheer: for I be-lieve God, that it shall be even as it was told me." Acts 27:25.

The lesson dealt with the folly of trusting self and the wisdom of trusting God, based on the Bible's description of the perilous journey of Paul from Malta to Rome.

Then, in the Life Lines for the day, I heard these pro-phetic words: "To save a person, many times it is necessary for God to allow his life to be completely shattered, his 'ship of life' destroyed. Then it is, that God reveals a refuge, an island in the midst of the stormy sea, to which he can cling for salvation."

The Debbie Fox Foundation was set up as just such an island in the stormy sea. It took shape very quickly. A five-thousand-dollar grant from the Hearst Foundation got things started. All those dollar bills and ten-dollar bills in letters from people around the country began piling up. Kind people volunteered to screen the hundreds of letters that were coming in from parents. They wrote back to the families, sent them medical reports, referred medical records of afflicted children to doctors and dentists. Doctors in Chattanooga and elsewhere reviewed the cases without charge when parents couldn't pay. In some cases the Foundation pays medical fees. But mostly it takes care of transportation of children and parents to hospitals and medical centers. Sometimes it works with Blue Cross–Blue Shield to cover deficits.

(For my first trip to Baltimore, the Chattanooga Helping Hand Club helped with our transportation. For all operations after that, the Foundation has paid our way to Baltimore or to Charlottesville and Mama's expenses while staying there.)

The Foundation's first case was a little girl from Kentucky who was born without a nose and with heavy ridges on her forehead that reached back into her brain. Her problem was too serious for her home-town doctors. The Foundation investigated her case carefully and then arranged for her to go to Johns Hopkins. Her church paid for some of her travel expenses. Dr. Edgerton, I heard later, made her a nose from surplus skin and cartilage in the facial area.

Soon after that, a fourteen-year-old girl from California

was getting facial repairs that she needed urgently, then a thirteen-year-old boy from Georgia, a seven-year-old boy from Iowa, a seventeen-year-old girl from Florida, a fourteen-year-old boy from Kentucky.

Up to the present time, the Foundation has referred hundreds of children, and some adults, too, for medical advice and played an active role in corrective surgery for more than thirty. Since he performed that first big operation on me, Dr. Edgerton and his team have done cranio-facial surgery on more than 150 children. Each year they try more difficult operations. Surgeons in other hospitals are also doing miracle operations.

I'm told there is no longer such a thing as a hopeless case, no matter how afflicted the child was at birth or how dreadful the damage done in an accident or fire. There is no need for children to hide in the shadows and lead blighted lives. They can go out and face the world, as I did.

# ∽18∾

# "Were You Scared, Daddy?"

School was going well. I was getting my little demon to quiet down. My face was becoming presentable. Now that she didn't have to stay home to take care of me, Mama was working in a school cafeteria. She enjoyed it a lot, and it meant a little more money coming in. That made life much easier all around. I began to think that maybe the worst was behind us.

It wasn't.

Daddy, you'll recall, worked as a spray painter at a company that made heavy-duty equipment. One day he saw two big cranes used for road work being put on a flatbed. Daddy said to his boss, "Don't put those cranes on a flatbed like that—they'll tip over." Nobody paid any attention to Daddy.

Daddy saw the cranes starting to fall, just as he'd predicted. He scrambled to get away, but he couldn't make it. His right leg was crushed under the huge, heavy steel.

"I don't see how in the world that crane kept from killing your father," one of the other men who worked there told us later.

Daddy was rushed to the hospital, into surgery, and they repaired his leg as best they could. It took a long time to heal; it was two years before he could really walk right, even longer before he could drive again.

I was frantic with worry for him. When I first went to see him in the hospital, he was so pale and tired-looking that I was really frightened. But he managed a smile for me, and even a little joke. "Hey there, baby," he said, "we've got things a little mixed up. You're the one who goes to the hospital and I do the visiting."

If he could joke, I knew, he was going to be all right. But it was a hard time for all of us. Blue Cross, Blue Shield, and Major Medical paid his bills. Mama and I visited him every day in the hospital. Then, when he got out, Mama took Daddy and me to her sister's house to stay for a while. Since Mama was gone all day at work and I was at school, we couldn't leave poor Daddy at home alone. It wasn't just bad enough that he had to go to some-body else's house to recuperate—I had to leave to go to the hospital while he was still sick.

My surgical appointments at Johns Hopkins were made many months in advance. If I canceled, it might be a year or more before another date could be set, and I'd get far behind in the work on my face. I was all for canceling. I didn't want to leave Daddy. He wouldn't hear of it.

"If you don't go, I'm not going to get well," he told me.

Well, that gave me no choice. He looked so pitiful, I hated to leave him. I told him, "I'll be home soon, Daddy, you just cool it until I get back."

So Mama, Mrs. Apple, and I made that long drive to Baltimore. Dr. Edgerton reconstructed the tip and septum of my nose, worked on my left cheek, and began straightening out my left eyebrow, which wandered all over my forehead. I don't remember much about that hospital stay —my mind wasn't on it. Daddy called me every night and told me to hurry home. I healed very quickly, and the doctors let me leave the hospital two days early. I'd sworn before, after that first awful flight, that I'd never take a plane again. But now I insisted on flying home—I couldn't wait another extra minute to see Daddy.

We got in at midnight, and I guess Daddy had been doing a lot of healing too while I was away, because when we touched down, he was at the airport in a wheelchair, pushed by Joyce. Of course our plane was a few minutes late, and he was all nerves worrying that maybe something had happened. He knew how I hated to fly. I was the first one off the plane. I practically slid down the steps of the ramp and raced to the terminal. I threw myself into his arms without tipping over the wheelchair, and he kept saying over and over, as if he couldn't believe it, "You're home, you're home."

"Of course I'm home," I told him. "I promised you I'd be right back." We all stayed at Joyce's that night. It was crowded in her house, but her husband, Marvin, and her boys, Mike and Jimmy, were happy to see us. It was a joyous reunion.

Once the bad part of 1970 was over, things began to get better. The people who were running the Debbie Fox Foundation decided to hold a banquet in Chattanooga to focus attention on their campaign to help children with cranio-facial problems. There was some talk for a while that I would appear at the banquet. I was excited at the idea. Then I began to worry. How would they feel about looking at me? I really, really wasn't ready to stand up in front of a banquet hall full of strangers. When the grown-ups decided I wasn't quite ready for a public appearance, I didn't make any fuss at all.

Then something amazing happened: The mayor of Chattanooga proclaimed the day of the banquet Debbie Fox Day. I couldn't believe it when I read it in the newspapers. I guess I began to believe only when I saw the ornate scroll with my name in big curvy letters.

More than four hundred people attended the banquet in the Silver Ballroom of Read House in Chattanooga. There were a lot of celebrities—the Sons of Song, a vocal group, actor Mike Mazursky, Doc Anderson, who is a famed mentalist, actor George Raft, Charles Wood, an industrialist from Alabama who lost most of his face in a fiery plane crash in World War II, actress Virginia Gray, John Shaw, an important oil man from Texas, and, of course, Dr. Edgerton.

I met some of the celebrities at a small reception the afternoon before the banquet. I wore a blue dress with red stripes, a small locket around my neck, and a new blond wig Mama had bought me. When Dr. Edgerton asked how I felt, I told him, "It's just like I said before—blondes have more fun." Everybody laughed at that. At the banquet they played a short tape I had made. On the tape I sang one of my favorite hymns: "Then sings my soul, my Saviour glad to be, how great Thou art, how great Thou art."

The banquet was a big success. It got a lot of publicity for the Foundation, and the people in Chattanooga who'd done so much for me finally had a chance to meet the famous Dr. Edgerton. At the banquet Daddy did something that made me very proud of him. Of course Mama and Daddy were there as guests, dressed up in their very best. They listened to all the speeches. And their hearts were full of gratitude. So full that Daddy just couldn't help himself.

After the other speeches, Daddy got to his feet with some difficulty (he was still recovering from his accident), and said to the people there, "I'm Debbie's father. And I want to express my appreciation from my heart for all the wonderful things you've done for Debbie. I've never spoken up like this before. But I have to tell you how much your kindness and generosity have meant to us. We'll never forget you."

Daddy was shaking when he sat down. Mama reached out and put her hand over his. The clapping went on and on after he finished. Madge Apple told me afterward, "Debbie, your father was beautiful. I cried, and I think just about everybody there had to pull out a handkerchief and wipe away a tear."

"Were you scared, Daddy?" I asked him later.

Daddy ruffled his hand through my wig—I have to tell you that didn't feel as good as having him ruffle it through my hair—and he said, "Debbie, I had to say my thanks for you. Next time, you'll say them for yourself. Is that a promise?"

"It is." I reached up and gave him a big kiss.

# ∾ 19 ∾

# *The Girl Who*
# *Found a Face*

About that time, I was outdoors one day looking at the ruin of my garden. Mama and I had been gardening together for several years—we planted flowers and vegetables, and I helped her take care of them. That year everything was a mess. The potatoes were rotting because of too much rain. A groundhog ate all the little cabbage plants. A hailstorm ruined the lettuce. The mail arrived while I was trying to decide whether or not to give up on the garden altogether.

Among the letters was a magazine—a copy of *Good Housekeeping* with an article about me entitled "The Girl Who Found a Face." It told, in briefer form, some of the story you've been reading here. Soon after that the *Reader's Digest* reprinted the article. Once more the mail began flooding in, just as it had after the big operation. There were gifts and good wishes for me, pathetic letters from parents seeking help for their children. This time those letters went to the Foundation, which answered them with helpful information and a brochure about its work. There were also many checks for the Foundation.

I never can get over the generosity of people and the way they respond to a total stranger when their hearts are touched. I recently looked at some of the letters I received then. I'd like to share a few with you.

One is just a sheet of notebook paper that says, "God Bless You."

On a small green card: "After reading the article, I couldn't resist helping. I haven't helped anyone less fortunate than me in a long time."

From Oregon: "Please accept my meager check for $5. I read about the marvelous, wonderful work of the Foundation. Words, except superlatives, fail me to say what I think this work will do for people. All good wishes to all concerned."

From Canada: "Enclosed please find a small donation for the Foundation. Our own son was born with bilateral cleft palate. He has been a delightful child and my heart aches for any of these children who must bear the burden of being 'different.' Debbie Fox is an inspiration to us all—it is truly the 'inner person' who matters, not the exterior shell."

On pretty blue paper with a gold design: "I am suffering from a facial problem called tic douloureux and I have been full of pain and self-pity until I read this article. Enclosed is a small donation to help a child."

A club of the wives of American servicemen in Holland sent a check for ten dollars. Several bowling clubs sent contributions. There was a gift from a woman whose nephew had lost three fingers in a boating accident. Many small gifts from retired people. At the office of the Foundation, a business-sized envelope was opened and out fell five one-hundred-dollar bills—no note, no name, no message.

From Wisconsin a woman wrote, "Our first little grandson was born with a harelip and cleft palate, also a cousin's little granddaughter with the same defect. We are very thankful for the fine surgery done on them—they are both beautiful babies now. Hope this will help Debbie and other needy children like her."

A sixteen-year-old girl sent a dollar and wrote, "I guess I took being a normal human being for granted. I never knew such defects existed. I am glad that you are trying to help these people. It's a wonderful cause."

From Canada on a small scrap of paper: "I am a fourteen-year-old girl. Tonight while I was babysitting I read about what your organization is trying to do. I decided to send my babysitting money."

From California: "In a minor way I have suffered from this type of abnormality and its psychological crippling. My particular problem could have been corrected, but was not acknowledged by an emotionally disturbed mother and an uncaring father. Good nutrition alone would have overcome the most devastating physical problems. Thanks for listening! The article helped me to understand myself a little better!"

That letter really got to me. It had never crossed my mind that I might have had parents who were uncaring and might have let me suffer a lifetime of deformity and misery. I let myself think what it would be like for me at fourteen if Mama and Daddy had left me as I was that day I first picked up the mirror and screamed in horror at my own face.

I walked into my bedroom with its pretty new white furniture and ruffled blue canopy and bedspread. It had been my big surprise present at Christmas. I looked at myself in the big mirror over the dresser. Yes, I had a big, beautiful mirror in my bedroom. The mirror was no longer my enemy. It was getting to be my friend. I saw smiling back at me a small teenage girl, with brown hair too short to be pretty. My eyes were blue-gray, the left one a little lower than the right one, the left lid still not right, with some leakiness around the left eye where the tear duct wasn't working properly. The eyebrows were still uphill and down. The nose was quite far from delicate or pretty. The mouth area was tight and pulled in. The lips were not well defined. The cheeks were a little sunken in spots because the bone structure was wrong underneath.

*But it was a face*—where there had been none before! A face that had been put there because my parents loved me, in spite of what I was. Because other people in Soddy and Chattanooga cared about me. Because a famous doctor dared to try an operation that had never been performed before. And, above all, because God's goodness reaches out, even to the fallen sparrow.

I could feel tears roll down my cheeks—tears of gratitude for the life I'd been given. I fell to my knees beside the bed, and with my wet face buried in the blue spread, I said, "Thank you, God, for making me whole again, bless Mama and Daddy and Dr. Edgerton and the people at the Foundation and all the suffering children. Thank you for making them well because so many people have loved me."

A few days after that Mrs. Huckaby told us all to wear nice clothes to school the next day. A photographer was coming to take our class picture. I put on my favorite—a purple jumper with a white blouse. Mrs. Huckaby and the photographer lined us up in two rows—the smaller children like Lisa, and most of those in wheelchairs, in front, the taller ones, like Janet and me, in back. There was a lot of kidding and pushing. Lamar kept asking, "Where's the birdie, where's the birdie?"

Lisa, always full of mischief, repeated, "Say cheese, say cheese!"

Mrs. Huckaby was busy saying, "All right, children, let's calm down."

I was doing a lot of hard thinking. Was it time, at last, for me to look forward into the camera? I'd made worlds of progress since that first backward-facing photograph with the school board people when I was in third grade. For the *Good Housekeeping* article, I'd had three pictures taken— one with Mrs. Huckaby in the classroom, one playing checkers with Daddy at home, and one going over homework with Mrs. Apple. All three showed the back of my head.

Was I ready to turn around?

No, I decided, I wasn't. Not yet. And without anybody telling me to, I faced away and let my eighth-grade class picture in June 1970 show me with the back of my blond wig and the back of my purple jumper to the camera.

It was the last picture in which I hid my face.

# ❧ 20 ❧

# *I Face the Camera*

In 1971 Dr. Edgerton moved from Baltimore to Charlottes-ville, Virginia, where he set up a cranio-facial department at the University of Virginia Hospital to take care of children with severe facial deformities. By this time, the risky eye-moving operation, which had been done on me with so much fear, was being performed almost routinely by Dr. Edgerton and other plastic surgeons. It was still a long, difficult operation and still required a team of top surgeons. But it was so perfected that there was almost no risk of blindness or death. Of all the hundreds of children who've had their eyes moved since I have, only one has died, and one other lost sight in one eye because of this type of surgery.

In June I went to Charlottesville to have work done on my lips and the walls of my nose. I was such a veteran of hospitals by then that it didn't bother me that I had to start over in a new one. In fact, I liked the hospital in Virginia right away. It's part of the beautiful campus of the University of Virginia. There are lawns and big old trees and handsome red-brick buildings, many with white columns in front and some designed by Thomas Jefferson. Everyone at the hospital was so friendly that I had no trouble feeling right at home. Don't misunderstand—I still hated the green operating-room suits as much as ever and it wasn't any fun at all getting ready for more surgery. It just had to be—that was why I accepted it.

This time Dr. Edgerton cut into the red part of my lower lip—the doctors call this the vermilion—and got ready to transfer some of it to my upper lip, which still had that too-tight cramped look that you see on the faces of people who've had a harelip fixed. At the same time, he filled in a depressed area in my left cheek where there was a small bone missing.

In July, I was back in Virginia again to have that tissue moved from my lower to my upper lip. It's impossible for me to tell you how he does it. It's been explained to me, but, don't forget, I have a terrible time just turning a hem, so don't expect me to understand how the plastic surgeons are able to interweave human skin so delicately that you hardly know it's been worked on. The stitches look bad at first, but soon the scar lines fade away. Then it's hard to remember what that part of the face was like before it was rebuilt.

That school year, my ninth, went along without much excitement. I think we were entitled to some peace and calm. Mama, Daddy, and I made the most of it. Mrs. Huckaby kept me on my toes about my lessons. I still wasn't having much success getting Lamar to pay attention to me, but at least I wasn't playing dumb tricks anymore.

I now had a little job at school. During lunch hour I set up a stand to sell school supplies—pencils, notebooks, erasers, and things—and the children from all of White Oak School, not just the handicapped children in my class, came to buy from me. The money went into a school fund for special treats. I got to know some of the children outside my class, from selling things to them and making change. Unfortunately, I couldn't really make friends with any of them; it was too hard to see them after school. They all lived near White Oak School; I had to ride a long way on my special bus. Remember, I was four hours on the bus every day, so I didn't have much extra time.

I rarely got to see Karen or Kathy, because they were at the Soddy-Daisy Junior High School and busy with all kinds of junior high activities. We still phoned each other once in a while, and I always liked to hear from them.

That was the Christmas when Daddy wanted to surprise me with a big stereo set in a handsome wooden cabinet. Up to then I'd been listening to my country music records —Loretta Lynn was my favorite—on a child's windup record player. I'd been hinting for a big stereo with good speakers, but I wasn't sure my hints were getting me anywhere.

One day just before Christmas, I got off the school bus and saw a large packing case on our porch. I poked at it until I'd figured out there was a stereo inside.

That night when Daddy came home, I asked him, "Daddy, whose stereo is that out on the porch?"

"Stereo on the porch? I don't know what you're talking about, baby."

"You know perfectly well there's a stereo out there."

"No I don't. Show me."

I opened the front door and, guess what, no stereo. "Well, there *was* one here," I told him. "I saw it myself."

He shrugged his shoulders and said, "It must have been a mistake. I guess the freight man took it away."

"Daddy," I told him, and I looked him right in the eye. "I *saw* a stereo on the porch, and I don't think it was a mistake."

Well, he kept up that "mistake" business right until Christmas Eve. Of course, he'd moved the stereo into a shed back of our house to hide it from me. It wasn't supposed to have been left where I could see it in the first place.

Christmas Day, when I found the stereo in the living room, I said to him, "Daddy, it's a funny thing. That freight man brought back the stereo he delivered by mistake and then took away. Do you think it's a mistake again?"

"Well, if it is, why don't you just enjoy it until he comes back to get it?"

That's how I got my beautiful stereo, and it was one of my best Christmases ever. Joyce had taken me to do my Christmas shopping. I'd bought scarves and gloves and handkerchiefs and bedroom slippers and things like that for Mama and Daddy, for Joyce, Marvin, Bobby, Janice, Ralph, my nephews, and some of my aunts, uncles, and cousins. Everyone came to our house for dinner. Mama baked a ham. I made a chocolate cake—Mama helped me some with the icing. We had a good time being together and feeling that at last our lives were getting better.

I knew things were really better when I posed with my class that spring for our ninth-grade class picture. I wore a tan jumper, and I looked straight at that camera, just as proud as I could be. Nobody told me to look forward or backward. I just knew I was ready to show my face. And from the way Mrs. Huckaby smiled at me, I knew she agreed.

Later, when I looked at the picture and saw my scarred

face and Lisa and Lamar and Alan in their wheelchairs and some of the others with their heavy leg braces, I remembered something Mrs. Apple had told me a long time before. Once when I was feeling sorry for myself, she'd said, "Debbie, we all have our handicaps, all of us."

"Not as bad as mine," I'd said.

"They may not *show* as much as yours," she'd answered in her quiet way. "But they may hurt more. Do you think you can ever know how much pain there is in someone else's heart?"

Well, my heart was full of happiness, not pain, when I took that photograph home. Daddy had a nice frame put on it, and he placed it on the mantel next to a color picture of Joyce that showed off her perfect features and made her look like a movie star.

"Now I can admire both my pretty girls," Daddy said. I tell you, my heart did a flip-flop when I heard that. Would anyone have thought fifteen years earlier, when I made my pathetic entrance into the world, that my picture and Joyce's would ever be up on the mantel side by side?

To think that my happiness was rippling out to others! Rippling far out, too, because at about this time a little girl named Faith Jacob came to Chattanooga all the way from India to have her face fixed. Her father, Devraj Jacob, was a technician for Air India. He'd read about me and the Foundation in a copy of *Reader's Digest,* which had been left in one of his company's planes. He wrote to the Foundation from India. Mrs. Apple replied, asking for more information about his daughter.

He almost didn't get her answer. For a long time no letter came, and he figured he was too far away for the Foundation to bother with his problem. One day he went back to visit the house his family had moved away from. There, scattered on the floor, were pieces of a torn-up letter from Chattanooga. Workmen in the house must have torn up the letter and thrown it away. He pieced it together, wrote again to the Foundation, and sent pictures of his daughter.

Faith had been born with a cleft palate and taken to London for surgery when she was eighteen months old. I guess the surgery didn't work very well, because her teeth came in very deformed. People in India must be even more uncharitable about deformity than they are here. When Faith went to school her teacher sent her home—she was too ugly.

"Go home and put your face in the sand," the teacher told her. Can you imagine that? Faith had been hidden away at home ever since.

Faith and her father came to Chattanooga, and there were many stories about them in the local newspapers. Dr. Phillips did some dental work for her. It greatly improved her appearance. The American plastic surgeons referred her to a plastic surgeon in India who could take care of her on a long-term basis.

It really thrilled me to think about a child coming halfway around the world to have her face improved. I was especially interested to hear that her father was a board member of the international organization Youth for Christ. From then on, I included little Faith Jacob in all my prayers.


In 1972 Dr. Edgerton did a doubleheader operation on me. One of the things he wanted to do was to get my eyes lined up better. Because my left eye was lower than my right, my eyes didn't look at all like a matched pair. "Let's see if we can't get these eyes to look more alike," he said to me before the operation.

We always had a little chat before I went into surgery, and he usually asked me what I wanted done most.

"What about my nose?" I asked him.

"We'll get to that, Debbie," he told me. "This time I want to work on your eyes. And we're going to give you a finger. What do you think of that?"

I held up my little stump. "I'm going to get a finger on this? Where's it coming from?"

He pointed to my right arm, near the elbow. "We'll find a little piece of bone in there you'll never miss and we'll move it to where it can do you some good."

"Well, hand," I said, looking at that useless little thing, "you're finally going to get fixed up."

"Not completely fixed, Debbie," the doctor warned me. "Don't expect too much. We're just going to give you one finger this time. Then maybe later on we'll use some of your toes to give you some more fingers."

I didn't want to hear anything about using my toes. It was okay to fix up the parts of me that were in bad shape. I certainly wanted that done. But I didn't even want to think about taking away toes to make fingers. My feet were perfect! I didn't want anyone cutting into them. Not even Dr. Edgerton.

The operation that time took five hours—the longest since my big one. First the doctors drilled through the bone between my eyes and threaded through the hole some tendonlike strips they got from my chest. They used these strips to anchor my eye openings and pull them into better alignment with each other. They also filled in a bald patch in my right eyebrow with some hair follicles transplanted from my scalp. Then they took an inch of bone out of my forearm, attached it to my little hand stump in the place where the pinky should be, and covered the bone with some skin from the stump end of my arm.

I went home from Virginia very excited about my hand. One finger wasn't much—but it was certainly better than none, and I looked forward to being able to do a lot more things for myself. Mama had instructions for changing the dressing on the finger. Both of us tried to take good care of the new little finger, but it just didn't seem to be right. After a while, it began looking red and angry. Then the bone seemed to be poking through the skin.

We went back to Charlottesville, and Dr. Edgerton shook his head unhappily.

"Debbie," he said, "I'm afraid we've got some trouble with this finger. And I don't think you'll like what we're going to have to do now." He explained that the skin placed over the little bone hadn't taken and the bone had become exposed. He said I was lucky I didn't get a bad infection. Now, to get proper skin coverage, he had to do what plastic surgeons call a two-stage flap. All the bits of skin and tissue that I'd had moved around on my face had come from close by and weren't very thick, so there wasn't any problem of getting a blood supply for them when they were put in a new position. The surgeon just had to take the skin from one place, put it in another, and sew it down.

This was different. The surgeon had to cut a flap of skin from my abdomen, but still leave it attached at one end so it would continue to have a blood supply. Then he had to attach the loose end of the skin to my finger. When the finger end of the skin had grown in and developed its own blood supply, he could cut the other end away from my abdomen.

It meant that for three weeks my right hand had to be tied to my stomach. I tell you, those three weeks were truly awful. I couldn't dress myself because I couldn't put my right arm into a sleeve. Mama had to drape clothes over me as best she could. I could sleep on my side or my back, but I had to be careful about not jolting my arm or pulling it away from my stomach. I couldn't go to school. I tried doing some of my homework, but I really wasn't in the mood for it. I think what I did most was complain. Poor Mama had to listen to a lot of wailing at that time.

Finally the three weeks were up. We went back to Virginia. Dr. Edgerton cut my finger away from my stomach and used the flap of skin that had once been on my abdomen to cover over my new pinky. He also made a kind of indentation at the thumb end of my stump, so that, while I didn't get a whole thumb, I got a slight one. As a result, my hand, if you could call it that, was a little more useful.

After all the trouble and pain, it was good to have a hand that worked at least a little. I could pick up things much better now. It was easier to cook. And there was a big new first: I could knit. Mrs. Huckaby had tried before to teach me to knit during arts-and-crafts time at school. I hadn't been able to do it. Now I could catch the knitting needle between the almost-thumb and the pinky. Mrs. Huckaby anchored it into that "hand" with an Ace bandage. Pretty soon I found out I didn't even need the elastic bandage. I just tucked the end of the needle between my right elbow and my right side and got along fine.

In January 1973 I finally got some more work done on my nose, both inside and out. Shaping the soft tissue of my nose was a slow job. Sometimes I thought of Dr. Edgerton as a sculptor working in slow motion. He'd refine the bridge a little. Six months later, he'd smooth out the nostrils. Some months after that, he'd fix up the side. Too bad there's no way to work with skin as quickly as with clay or even marble. It was really a miracle to work in skin and flesh at all, and my nose was a terrible challenge, even to a miracle-working plastic surgeon. Imagine: He had to start with hardly more than a hole and build up, piece by piece, that amazing organ we call a nose.

As Mama and I went back and forth to Charlottesville on the bus—sometimes I felt we were almost commuters—I had lots of time to think about my gratitude to the Debbie Fox Foundation, which paid for our transportation and Mama's room at a boarding house just a short walk across the campus from the hospital. Then there was my gratitude to Dr. Edgerton and his colleagues. Even after all this time, there were still no bills from him.

Someone once told me that if there had been a charge for that first big operation, it would have come to somewhere between $50,000 and $100,000. I wouldn't doubt it—with six surgeons, their assistants, nurses, and technicians and use of the operating room for more than twelve hours. The operations after that would normally cost from $1,000 to maybe $10,000 each. Then there were all the earlier, less extensive operations by Dr. Barnwell and by the doctor in Atlanta.

How much would the surgical fees have been all together if we'd been paying? It's hard to say. Maybe half a million dollars. Maybe more, maybe less. It wouldn't matter; we couldn't possibly have paid it. Blue Shield wouldn't have made even a dent in such charges. I often wonder how other families of children with severe defects manage when they get surgical bills bigger than the whole family's income for the year!

I didn't know anything then about the economics of medical care. I did know that those families needed help— lots of help.

# ∾ 21 ∾

# *The Shadow of Death*

February 23, 1973, started like any other day. It ended as the worst of my life.

It was a Friday. Daddy came home from work at about four thirty, and Mama, Daddy, and I sat down to supper. Daddy had brought home his heavy-work clothes, and after supper Mama put them into the washer. I was stretched out on the couch with the beginning of flu or a cold. In recent years my after-supper ride with Daddy had changed into an after-supper shopping trip. That night I told him, "You go to the store, I'll do the dishes."

"Anything in particular you want, baby?" he asked.

"No, nothing special."

He was gone about an hour. When he came back with groceries and other things, I asked, "Hey, did you bring me a Coke?"

He hadn't, so he went off to a nearby stand to get us Cokes. He was gone only about five minutes.

There was a good movie on TV that night. We all sat

and watched it until about nine thirty. Then we went to bed. At ten I heard odd noises. Daddy was having trouble getting his breath.

Mama called to me, "Debbie, you better get up. Daddy's sick."

I went into their bedroom, and Daddy was all drawn into himself. It looked as if he was having a convulsion. Mama and I pulled him to the side of the bed and washed his face. Then I tried to phone Brother Hickman, who lived right across the way. He was out of town. I told Mrs. Hickman to come over; something was wrong. She threw a coat over her nightclothes and ran over. In the meantime, her son called for the ambulance. It came in less than ten minutes. But Daddy was already gone.

Daddy was dead. My beloved daddy had been taken from us.


Brother Hickman arrived just as the ambulance turned in. He thought something was wrong with me and rushed over to ask what the trouble was. I couldn't speak. I just pointed to Daddy. The preacher turned and saw him on the bed. I've never seen anyone get so white. Poor man—I felt so sorry for him that I almost forgot about myself and how stunned I was. The preacher said a prayer for Daddy and made Mama and me promise that we'd call him if we needed him during the night.

The ambulance took Daddy away. Mama and I stood in the doorway looking after the ambulance as it turned up the road and disappeared.

I looked up at the sky over Soddy Mountain and listened for the sound of the creek. There were no stars, just darkness and the soft ripple of the creek. My eyes were all teared over as I looked up to heaven and prayed, "Dear God, bless Daddy and take him to You."

Everybody had liked Daddy. He was a good man: kind, understanding, and gentle. He'd only gone through grade school, but he was a lot smarter than many educated people. The next morning so many people sent messages and flowers and food to the house that Mama and I had trouble getting away to Williamson's Funeral Home in Daisy to make arrangements. We stayed at the funeral home most of Saturday. His friends came from everywhere to pay their respects—from work, from church, from the neighborhood, from the town where he was born.

On Sunday afternoon the church was so filled for the funeral service that people were standing out in the hall. Carnations, chrysanthemums, and roses were massed everywhere. Daddy was in an open casket, and when the family went up to the casket to have a last look and tell him goodbye, my younger brother Ralph, who was twenty-seven, took it so hard that he was almost out of his mind with grief. Shaking and sobbing, he tried to pull Daddy out of the casket—to bring him back, to make him live again.

Someone screamed when Ralph took hold of Daddy, and then the funeral director was at his side, saying, "All right now, take it easy."

We finally got Ralph calmed down. He began to realize fully that Daddy was gone. We all knew now that we'd never see him again. We felt we'd all be lost without him. During the service Brother Hickman's powerful voice prayed for Daddy and brought words of comfort to the rest of us, reminding us that Daddy was in heaven, that he was saved, that he now walked with Jesus.

I said my own special prayer of thanks that Daddy had gone back to church with me and found God again, for now he would surely have peace. The quartet sang two songs, "City of God" and "Oh, What a Happy Day."

We all went to the graveyard out on the Dayton Pike. It had been snowing. There was still snow on the ground, but now it was raining. Sadly, we said our last farewells. I was choking with sorrow. But even at that terrible moment I knew I would have to accept the loss of this good man who had given me so much love and joy. I had been his baby. Now I was not a baby anymore. There was just going to be Mama and me in the house.

It would never be the same again. I would never be loved that way, ever again.


That night, after all the others had gone, Mama and I sat at the table in the dining room and had a cup of tea before going to bed. "It's just us now," Mama said. "Just the two of us. Do you think we can make it?"

I looked up at the picture of Jesus on the wall. In the picture, His head is bent and He looks gentle and thoughtful and very young.

"Not just the two of us, Mama," I said. "We're not alone. We're never alone."

I've always believed that.

# ∾22∾

# "These Are the Gifts I Give to Thee"

Less than a month after Daddy's death, the Debbie Fox Foundation was to hold its second banquet. This time I was scheduled to be one of the honored guests. I'd known about the banquet before Daddy died, and I'd been thrilled when I was invited. I'd already begun getting ready for my part in the program. It was a very important event: I was to make my public debut. I'd never before stood and shown my face to a banquet hall full of people. Should I go ahead and appear at the banquet while we were still so full of grief?

"Mama, what do you think?" I asked her.

"I don't know that I can help you, Debbie," she said. "I think your daddy would have wanted you to go ahead. But you might not feel right about doing it so soon. I just don't know."

"What would you do, Mama?"

She shook her head. "I don't rightly know."

I decided I'd ask Mrs. Apple. She didn't help me much either. She said, "Debbie, we'd understand and accept your decision either way. We're all looking forward to having you on the program, and we'd like for you to be with us. If you feel it's too soon, just tell us and we'll understand."

I could see that Mama and Mrs. Apple both wanted me to make up my own mind.

I was thinking very hard about what I should do. One day while I was waiting for the school bus I looked at the lilac bush in our yard, just getting ready to break into bloom. I saw a shadow whirl by one of the almost-open blossoms. I thought to myself—it can't be a hummingbird yet, it's too soon, spring is hardly here. Then I remembered one of my poems. It was a long time since I'd written a poem—not since I'd gone to school and come out of my lonesomeness. Now I remembered my poem about the lilac bush. This is the way it went:

> The flowers of lilacs are so sweet
> It seems a special treat
> To walk in the early spring,
> Its fragrance the breezes bring.
>
> It has a nice, scented bloom
> And the bees will have to make room
> For the hummingbird to tarry
> Awhile and food home to carry.
>
> It blooms the earliest in the spring
> Breaks the winter monotony and spring it
>         brings.
> It makes us feel chipper and gay
> When the joy of spring we display.
>
> It helps us to sit and discern
> All the things God had concern
> All the ways He decorated this place
> Which seems to us such a big space.

It had been one of Daddy's favorites. When he first read it, he'd said to me, "Baby, that's a real pretty poem. You can do anything. You can do just anything you put your mind to."

Now I knew I had to go ahead and do things. I had to go out and find my place in the world. Daddy had wanted the world for me, God's whole big place. Now it was up to me. I told Mrs. Apple I'd be proud to be at the banquet and take part in the program.

Pickett's store in Chattanooga gave me my dress for the banquet. I chose a beautiful gown with a dark blue front and long pink sleeves. It was my first long dress and I was delighted with it. Sylvia Freedman at the Twin Cities Beauty Shop did my hair in lovely soft waves. I had a corsage at the beginning of the evening, but somehow it got lost in all the excitement.

When I walked into the Silver Ballroom at the Read House, the sight nearly took my breath away. The room itself looks like something in a castle in a fairy tale. It is a huge, high-ceilinged room with many tall mirrors. Soft blue-green velvet draperies are pulled back to the sides of the mirrors. The walls are a blue-green damask. Great crystal chandeliers hang from the ceiling, which is decorated with silvery filigree, and there are silvery light sconces on the walls. All that light from the chandeliers and the sconces fell on tables set with beautiful china and silver and flowers, and on all the people in evening cothes.

And all those people, most of them strangers to me, smiled at me! Some of them came over and introduced themselves or took my hand and said, "You must be Debbie, we're so proud of you."

I tell you, I felt like a movie star. Was I nervous or frightened about standing up before all those important people? I'll have to confess that I really wasn't. I was proud to show my face. Even though it was still not perfect, I wanted people to see what had been done for me. I had rehearsed what I had to do. I was ready.

I gave the invocation for the dinner. I stood up at my place at the long dais. I could feel those hundreds of eyes upon me. I held my head high and looked back at the audience so they could all get a good look at me.

I bent my head and prayed aloud, "My Father in heaven, thank You for loving us all, for being so kind to us and through so many people. May we return the love of Jesus, I pray in His name, Amen."

There was a hush as I sat down. Dr. John M. Higgerson, a prominent doctor in Chattanooga, presided at the banquet. Dr. McConnell, the superintendent of schools, was the toastmaster. I felt very indebted to Dr. McConnell, partly because the Hamilton County schools had done so much for me and also because his interest in my problems had made it possible for Mrs. Apple to take so much time off from her work to travel with me on my early hospital visits and to be around whenever I needed her.

The high-school chorus sang a song. John Germ, the president of the Debbie Fox Foundation for Cranio-facial Disorders, as it was now named, made a short speech. Dr. Edgerton gave the main talk of the evening. It was called "New Horizons in the Treatment of Children with Cranio-facial and Skull Deformities."

I learned for the first time, when he spoke, that he had made a ten-minute film of my operations. It had won the surgical-film award of the year at the American College of Surgeons' Clinical Congress. He recounted the history of my medical problems and told of the risks in changing the position of my eyes.

"At the time," he said, "there were no reports in English medical literature of correction by surgery of this type of deformity. It was the patient herself who asked for the chance to take the surgical risk of major cranio-facial surgery. In doing so, she became a medical pioneer and opened many doors for other children in the years ahead.

"Needless to say, when the Debbie Fox Foundation asked me to serve on their board of directors, I was honored and pleased. The new hope generated by these early surgical successes encouraged me and other members of my team to devote more time and study to the correction of these major defects in children. We decided to develop a special multidiscipline team of plastic surgeon, neurosurgeon, opthalmologist, speech and hearing experts, social workers, and others who might provide complete medical care for these children. In certain cases these services have been provided without professional fees for families unable to afford medical care.

"We now know that most of these children indeed have normal intelligence and most can withstand the long eight-to fourteen-hour operations required for correction. Most of them can then be given a genuine hope for a normal life and presentable appearance. The risk has proved to be not great in comparison with the gain. At Charlottesville I was offered full department status and the opportunity to develop a cranio-facial center with services for children from distant states. Subsequently we have carried out over twenty-eight major reconstructions in children with orbital hyperteleorisms similar to Debbie's. Each child presented a different problem, and new surgical principles have emerged from each operation. Several of the children were referred through the Debbie Fox Foundation. Plastic surgeons in Charlottesville and Chattanooga have thus begun developing a sense of fraternity in this new enterprise."

We were all deeply moved by Dr. Edgerton's words. Then came my big moment. For weeks I had been rehearsing a special song with Mrs. Ann Patton, a prominent woman in Chattanooga, the wife of a physician and a board member of the Foundation. Mrs. Patton has traveled all over the world as a harpist and singer. She has a beautiful golden harp which she plays with a special joy, for she is black and for a while she had been refused harp lessons because of her color.

Mrs. Patton took her place at the harp. I walked to her side, feeling my long dress rustle against my legs. We waited for absolute quiet. Then I sang a beautiful song, "Gifts." It was a poem by Marion Peck, who is a well-known writer in Tennessee, and set to music by Mrs. Patton. Here are the words:

I saw a child who could not see.
Dear Lord, I cried, why should this be?
Never to look on the face of love . . .
Or the sun . . . or the trees . . . or the stars above . . .
Always to walk through eternal night . . .
Never to know the comfort of light?

I saw a child who no step could take.
Dear Lord, I cried, my heart doth ache!
Never to sand up straight or tall . . .
This child, though he tries, can only fall . . .
What did he do to be made this way?
'Tis every child's right to run and play!

I saw a child who could speak no word.
Isn't this cruel, I cried, Dear Lord?
And he cannot hear, on top of it all . . .
Never can hear his mother call . . .
Never to speak . . . never to hear . . .
I turned my face to hide a tear.

I saw a child who could not reason.
Locked away for his whole life's season . . .
Locked away and all unaware

Of the world of the mind that others share.
Dear Lord, I said, why this affliction?
Denying this child the mind's benediction?

Then spoke a voice through my heart to me:

"THESE ARE THE GIFTS I GIVE TO THEE

That little child will take your hand
And lead you through this weary land . . .
And show you wonders beyond compare . . .
And beauty you never dreamed was there.
And that little boy so crippled and lame
Has courage, My Son, that would put yours to shame.
He'll teach you to walk if you'll just hold his hand . . .
To walk upright with your fellow man.
And the little girl there whose tongue could not speak . . .
Can teach you the eloquence of tears on a cheek.
You'll learn that the heart can volumes tell
If it's warmed by love and care as well.
And that mindless one . . . Hold the cup to his lips
And you'll find that God's love you also will sip.
Perhaps your mind's closed, but you hold the key
And when you open to him you open to Me."

Dear Lord, I said, I humbly thank Thee.
I'll live for these children . . . who are Your Gifts to me.

When the song was finished and before the applause could begin, I suddenly found myself standing and holding up my hand, my good hand, for silence. Almost as if I were somebody else, I heard myself say to that surprised audience, "I want to thank all the people of Chattanooga and the surrounding areas for the help you've given me and other children. You've taken our hands and opened doors for us and shown us brighter skies. We are where we are today by your help. I thank you with all my heart."

The applause came crashing down, and I sat down, all weak and surprised. I had amazed even myself with my words. I hadn't intended to speak out like that. But in a way, maybe, I had after all. Naturally, I'd been thinking a lot about Daddy that day, wishing and wishing that he'd lived long enough to see all the honor that was coming to me at the banquet. As I thought about him, I remembered how he'd gotten up to thank the people at the earlier banquet. I also remembered that he'd said to me afterward, "Next time, baby, you're going to thank all those people yourself. Is that a promise?"

Yes, it was a promise. And I'd kept my word to Daddy. I'd stood up and said "Thank you," just as he had.

Mrs. Patton and I had become good friends during the time we'd rehearsed together for the banquet. She lives in a lovely mansion on Missionary Ridge, overlooking all of Chattanooga. We'd had quite a bit of time to talk when Mama drove me there to practice with her. She'd worked with me very carefully and patiently to get the timing of my singing and the harp just right.

One day she'd said to me, "You know, Debbie, when I listen to you sing with so much feeling, I can't help thinking of all the children who are hidden away because of their defects. Do you realize how much you've done for them, to bring them out and let them make their contribution to the world?"

I was very touched and told her, "Mrs. Patton, I'm going to pray for you." I added, sort of mischievously, "You know, when I pray for somebody, something good happens to them." I didn't mean it seriously—it was just something I said.

After the banquet, Mrs. Patton came and took my hand,

and she said, "Do you remember what you said about praying for me?"

"Yes, I remember."

"Well, now I have to thank you. Because the good has happened to me. You've given the world a chance to see me, not just as a concert artist, but as a person with a part to play in making hurt children happy. You were right—God has blessed me through you. And God is going to keep on blessing you as you help other boys and girls to find their talent."

I think I did the right thing when I went to the banquet. It was what Daddy would have wanted me to do.

# ∾ 23 ∾

# *Living in the Present*

For quite a while Mama and I were so lonesome for Daddy that we hardly knew what to do. That fall Mama lost her mother, my grandmother. It was another real blow for her. Think of it—Mama lost her husband and her mother in the same year. In December Mama had to go into the hospital for a hysterectomy. The next July she was in the hospital again, for a gall bladder operation. It just wasn't our time.

When she had her hysterectomy, I had to go to Virginia for surgery on my cheek and mouth. Joyce went with me that time. You can imagine the state we were both in, worrying about Mama. The same thing happened in July, when Mama went to the hospital again—I had to go again, too. One of the things they did for me that time was to work on my tear ducts. It seemed like a bad joke. The doctors were making it easier for me to cry. And I had plenty to cry about.

But Mama soon got better and was able to go back to her job at the Kayser Roth plant not far from where we live. She inspects and folds the hose before they are packed. She has lots of friends where she works, and I think she likes it there.

Guess what got me through the hard times of that year—driving!

For the longest time, what I wanted most was to be able to drive. Driving around in the back seat with Daddy had been the biggest pleasure of my childhood. But when I got up into my teens I wanted to be able to drive myself. I'd watched how Daddy used his hands and feet, and I'd memorized most of the steps of driving. I was sure I could do it, even with my bad hand. The stump was good enough to use to steady the wheel. I could shift with the other hand.

The year before Daddy died he began taking me to practice driving on the ball field below our house. He didn't want me out on the road causing trouble. I guess he thought I'd be safe on the field. The grass was muddy and bumpy, and I had a terrible time. Besides, I think maybe men don't know how to explain things to you. Or maybe daddies aren't good at teaching their children. He got me so nervous that I had to say to him, "Daddy, you've just got to quit fussing!"

"I'm not fussing, Debbie, I'm just trying to show you," he'd say.

"But when you show me over and over, that's fussing."

"It's not my fault if you don't get it. Here's how you do it," and he told me all over again.

He kept getting after me, especially when I tried backing up. That was the worst of all. We went driving together quite a number of times, but I didn't make too much progress.

The winter after he died I was nearly eighteen and felt

I really needed to drive so I could go places by myself. I signed up with the Hamman Driving School for the Handicapped. I made the decision to do it while Mama was in the hospital for her hysterectomy. I didn't want to be stuck way out in the country and have to wait for Joyce or someone to drive me to the store or to the hospital.

I had my first lesson the day of her surgery. "Guess what?" I told Mama the next day. "I took my first driving lesson yesterday."

"How did it go?"

"No complaints."

The teacher was very patient with me. After Mama got home, I hated to leave her, but I did manage to get away for all eight lessons. I made good progress. After I'd finished the lessons, my learner's permit was going to run out on the next Sunday. I went to the Highway Patrol to have it renewed. The trainer, a nice woman, said to me, "Why don't you take your test right now? Then you won't need a renewal, you can get a real license."

I was floored. I wasn't ready, in my mind, to take the test. But I had nothing to lose, so I hopped into the car with the woman. She really put me through my paces: "Pull up to a stop. Turn left into that dirt road. Back in here. Try your blinkers. Park in front of that truck. Make a U-turn." I did it all without any serious mistakes.

"You've got your license," she said.

"Oh goody," I said.

Now I was free in a whole new way. I don't do a lot of driving these days, because Mama has to take the car to work. But I like knowing I can be off on my own any time I want. I want to have a car of my own someday and just follow the roads wherever they lead. I don't exactly think of myself as an easy rider—but maybe there's some gypsy in me after all.

I guess I haven't told you much about my life in high school—I've been so busy talking about my operations and all our other troubles. When I went into tenth grade, there was a lot of discussion between Mama and Mrs. Apple about whether I should stay in my special little class for the handicapped or go to the regular Soddy-Daisy High School. Academically, they said, I was ready for the big high school. They felt I could easily keep up with my classes. Even with the problem of going off to Virginia for two or three weeks at a time for more surgery, they thought I could manage.

Socially and emotionally, though, they didn't think I was ready. There's a lot of dating and social life in our high school; by the time most of the boys and girls graduate, they're just about paired off for life. Many get married right after high school. Even the graduates who go away to college often marry the person they were going steady with in high school.

Mama and Mrs. Apple worried that I'd feel bad about all the dating and flirting that would be going on around me if I were left out. And they were very much afraid I'd be left out because of the way my face looked. I remember one day Mrs. Apple said to me, "Debbie, you're just beautiful inside. But I'm afraid teenage boys aren't much at finding beauty inside."

Mrs. Huckaby also worried about the kind of time I'd have at high school. Once her daughter had brought a friend to help entertain at a party for our handicapped class. The girls were going to do a baton act for us. Mrs. Huckaby's daughter had warned her friend that we all didn't look so great. I suspect she particularly warned her friend about the way I looked.

Well, her friend could hardly get through their act— she was so upset by my face. As soon as she was finished, she just flew out of the room. Mrs. Huckaby understood that looking at deformity is harder for some people than for others. The other grown-ups knew that too. Rather than take the chance that high school might be a nightmare for me, they decided I should stay with my little group.

None of us will ever know whether that was the best decision.

What I gained was the security of staying with my own small group. It was also easier for me to be absent because of surgery and healing.

What I lost was the chance to make more friends and learn more about getting along with people my own age. I also lost all the fun of high school—pep rallies, homecoming games, junior and senior proms, the crowning of the carnival queen and the class beauties, and the recognition-day awards.

Would I have felt left out at all those activities? Would I have been miserable if I weren't popular, weren't invited, weren't chosen?

I don't know.

I do know we had some good times at our little school. We were no longer at White Oak School, but at the new Dawn School, where children with handicaps and disabilities have three beautifully equipped classrooms set around a little quadrangle and a special administration building with offices and physical therapy equipment. Dawn School is in a rural area, on a high hillside with a wonderful view down into the valley. It is the result of some more of Mrs. Apple's schemings. After the special classroom for the handicapped had been set up, when I entered the eighth grade, she wanted more and better facilities.

Federal and state governments were beginning to take

more notice of the handicapped about this time. More money was becoming available. So the Dawn School began. It was the first step in developing a wide range of services for children and adults with disabilities. I'm told it's one of the finest institutions of its kind in the country.

When it opened, the *Chattanooga Times* quoted the president of the local Easter Seal region: "The darkest night there can be is when a handicapped child is kept at home with no place to go and the child's parents must constantly be in attendance." The article then went on to say that the school grew out of the problems of Debbie Fox.

Maybe so. To a much greater extent, it grew out of the dreams of Madge Apple to bring more children and adults out of the shadows of their handicaps. Most people think the name of the school means a new dawn for those with special problems. It does. It also means something else. Dawn was Mrs. Apple's maiden name, and naming the school for her is a tribute she certainly deserves for a life spent working for the most needful of children.

During the years I was at Dawn School, I got over to the Soddy-Daisy High School, a big brick building on the Dayton Pike, from time to time. It was always fun to see the boys and girls I'd known only by their voices from my telephone hookup. Sometimes we'd talk for a few minutes and ask each other how we were doing. Kathy and Karen were always happy to see me. I went shopping with them a couple of times after school.

There were some very good-looking fellows on the football team and the basketball team. But I could only admire them from a distance. I wasn't in their world. I wasn't one of the pretty girls who climbed into a car with those broad-shouldered boys and drove away with a little extra gunning of the engine, just to show off.

Sometimes I'd look after them longingly, when the couples drove off together, or strolled by arm in arm. I very much wanted for myself this closeness to another person. I wanted it very deeply.

I often daydreamed of having one of those good-looking guys come up to me and say, "You know, Debbie, I've had my eye on you and I'd really like to know you better. Can we get together for a Coke after the coach lets me out of practice?"

Sometimes my dreams got even grander. When the senior prom was coming up, I knew the girls were very excited about their dates and the dresses they'd be wearing. I closed my eyes and saw myself in a full-skirted rose-colored dress with soft folds around the top. I saw a tall, handsome boy in a white tuxedo with a black band all around the edge of his jacket—that was the fashion that year—pinning a corsage of gardenias to the shoulder of my dress.

"Hold still, Debbie," he was saying. "If you wriggle, I might stick you with this pin." But I was so eager to see how the flowers looked that I kept wriggling. "Debbie!" he sort of growled to make me stop. I looked up at him. He faded away. My daydream ended. There was no good-looking boy, no date, no prom for me.

At least I had a good excuse for not going to the prom. I had another kind of date—in Charlottesville for more surgery on my nose and palate.


The high-school years sped by and suddenly came to an end. It was June 1974 and I was graduating. I was the only one in our special class to graduate that June. Janet had graduated the year before. I should tell you the good news that she recovered from her illness, married soon after graduation, and now has two lovely children.

Lamar, the boy in the wheelchair I wanted so much to be my special friend, had begun having more and more trouble with his muscular dystrophy. After a while he couldn't feed himself. I remember how angry he got at his hands when they wouldn't do what he wanted them to. I knew just how he felt. I guess we all did; we all had parts of our body that we could not make do what we wanted. Lamar got worse and died. A group of us sang at his funeral. Later the library at the Dawn School was named in his memory.

One of the boys with cerebral palsy got well enough to leave our class to go to the regular high school. He even made it to the wrestling team. One of the girls with a spinal injury lost both her parents and had a terrible time for a while. But now she works in a photo studio and drives her own car. Little Lisa, I think, is still in school.

The Soddy-Daisy High School holds its graduations in the Tivoli Theatre in Chattanooga, one of those ornate movie palaces left over from the 1930s. There were three hundred of us in cap and gown. The graduating class was seated down front in the auditorium, our parents and friends at the back of the hall.

There were a lot of speeches which I don't remember, people telling us to go out into the world and make our mark. Then we lined up on the big stage and were called up by name, one by one. I guess a lot of people knew about me, because there was a burst of applause when my name was called. I walked up and got my diploma. I felt very happy about it, because I'd worked hard for it for a long time. I'd achieved something I'd always wanted. I was also feeling sad and low, because Daddy was not there to share this important moment. I was thinking so hard about Daddy that I was almost in tears when I took my diploma in my hand.

Mama was there and my brother Bobby and his wife, Janice. Joyce was working and couldn't get away. We didn't have a celebration; we couldn't get ourselves to do anything festive without Daddy. Mama and I went home, and over coffee and cookies in the living room we talked about that first day so long ago when I stood on the porch with my pencil box and notebook under my arm and my hair streaming down my back—not knowing what my face looked like; not knowing that I hardly had a face—and watched my teacher come up the walk to start my first day of school.

Then we turned the pages of the *Trojan*, my high-school yearbook for the class of 1974. The book was a record of all the things I hadn't done—marching band, daisy chain, football rallies, teams, clubs, queens. Mama and I looked at the Carnival Queen, the Calendar Girls for each month, Mr. and Miss Soddy-Daisy and their Court of four couples, the girls in long dresses, the boys in tuxedos, all smiling.

There was my friend Karen McRee, the Homecoming Queen, with that special, sweet smile of hers lighting up her face. My other friend, Kathy Smith, was the editor of the yearbook. She wrote as her message to her classmates, "Just remember that even though it's nice to lose yourself in memories, we must live in the present so that we may be ready for the future."

I turned to look at my picture. Like all the other girls, I was wearing a black drapery across my shoulders that made a wide V in front. My nose still looked oversized. You could see the scarring around my eyes. My eyebrows were marching in all directions.

But there I was, looking straight out at the camera, proud to show my face, even with its scars. "My picture doesn't look too bad, does it?" I asked Mama.

"Debbie, I think you look real good," she said. Mama was quiet for a while. Then she asked, "Do you feel bad that you missed out on all those good times?"

I let the glossy pages slide through my fingers. "I guess

I missed a lot, Mama, but, you know, it's over for the others, too."

Suddenly I knew Kathy was right. Maybe my classmates had stored up more happy memories of their high-school days than I had. Maybe they'd had fun and romance and excitement, which had passed me by. But now these good times were all just memories for everyone. They were the past. Now our business was the present—and the future.

I closed the yearbook. But before I put it away, I reached into the drawer where I kept old papers and photographs. I pulled out a typed sheet of paper and folded it once before I slipped it into the yearbook.

"What's that, Debbie?" Mama asked.

"It's a poem—one of the poems I wrote years ago."

Mama didn't ask, but I knew she was wondering why I was putting it in the yearbook.

"The other girls had the corsages they wore at the prom," I explained. "I heard them talking about pressing a flower from the corsage in their yearbook. So I'll press this poem about our mimosa tree. I've always loved its flowers."

Mama and I read the poem together.

## A Pink Mimosa

A mimosa tree is a sight to behold,
Its shade is like pure gold
As I sit in its shade
Until the sun away will fade.

The flowers are like powder puffs
Wadded up in great big tufts.
The color is a baby pink,
It's really softer than you think.

Its leaves are like the web of a spider,
Between each part is a divider.
When touched, it closes together,
As if to remain that way forever.

They will grow side by side
Or set apart very wide.
They survive in any weather
As water from the soil they gather.


I picked up the teacups. "I'll wash these, Mama," I said. "School's over and I can sleep late tomorrow. But you have to get up early."

Mama gets up at five every workday, including Saturday.

# ∽24∽

# *I Have Been Set Free*

The years since my graduation from high school have gone by both slowly and quickly. Many of the days drag. I am alone a lot, keeping house for Mama and me. I read, watch television, walk down the street to visit Grace May, who runs a flower shop in her house. Grace and I exchange the news of the neighborhood—new babies, sicknesses, recoveries, an addition on someone's house, a lost job, a cousin arriving from out of town.

When Mama gets home with the car, we do the marketing, drive around a little, have supper together, watch television, go to bed early. Not a great deal happens. We go to church together on Sunday. Sunday nights I sing with the fifty-voice choir at our church. One summer I went on a camping trip out west with Joyce and Marvin and my two nephews. That was fun, and I began to get the travel bug. One of these days I really have to see the world. Right now it's a quiet, close-to-home time in my life.

I look around, and suddenly it's Christmas again. Or

the lilacs are in bloom again. Or, even more likely in my case, it's time to go back to the hospital again. Each year there have been two and sometimes three trips to Charlottesville for more of that slow, patient repair work on my face.

That's why I've had to put off choosing a field of work and settling into a job. Who wants a new, inexperienced employee who has to keep taking off for the hospital and then needs weeks of recovery for the swelling to go down and the black and blue to disappear?

One summer I worked at the Siskind Foundation in Chattanooga, where crippled children go for surgery and rehabilitation. I diapered the small babies, read to the younger children, played cards with some of the older ones. The nurses said I was really good at comforting the little ones who cried and missed their mothers. Should anybody be surprised that I'm good at that? I don't think so.

For a while I went to a rehabilitation program, but I found it wasn't teaching me any real skill that I could use in an office or in work with the handicapped. Besides, the program had hardly gotten started when I had to leave for the hospital.

Not long after graduation from high school I went back to Charlottesville for one of the worst of my operations— the worst if you don't count the big one. This time Dr. Edgerton did a lot of adjusting inside my mouth, working on my palates and in the area of my pharynx. I had severe pain that time and some heavy bleeding that really scared me. The idea was to make my speech clearer, and that's what the operation did.

The operation after that was a complicated one to shore up the lower part of my face. As I've told you, the bony structure of my whole face is badly messed up, and Dr. Edgerton seems to be completely rebuilding it from the inside out. He has to get the bones right before he can correct the contours in my flesh and skin. This time, one team of doctors opened my chest wall and took out two ribs. Another team made a tunnel inside the lower part of my face, threaded parts of those ribs through the tunnel, and wired them in place to give better support to my cheeks and firmness to my upper jaw.

By the time of that operation, Dr. Edgerton and his team had taken some of my ribs and put them in my face, some tendons from my chest and strung them between my eyes, and a bone from my arm and attached it to my hand. One day, when Dr. Edgerton came into my room and asked me, "Debbie, how're you doing today?" I said to him, "This has to stop—I'm getting too mixed up, the way you're moving parts of me around."

He smiled and tipped back my head to get a good look at my face. "I'd say you're doing fine, even with the parts mixed up."

Something he'd said some time earlier still bothered me. "You're not really thinking about moving some of my toes around, are you?" I asked him.

"Yes, I am, Debbie. I think you'll be much better off if you have a workable hand. Maybe we'll talk about that the next time you're here."

I didn't tell him then, but cutting off my toes was something I didn't want to talk about—ever. Much as I wanted a hand that would really work for me, I didn't want it badly enough to give up any of my toes.

My feet are like other people's feet—normal, with ten perfect toes. Little enough of me was normal at birth. How could I even begin to think of letting a doctor, even my beloved Dr. Edgerton, cut into my normal feet and

have them come out deformed? I couldn't. I didn't want
to hear any more about making fingers out of toes. I put
the idea out of my mind.

I've had so much surgery—I think the total is now fifty-
eight operations— that many of my trips to the hospital
have blurred together in my mind. There have been so
many rooms, roommates, nurses, operating rooms that I
can't really keep them apart. I just remember bits and
pieces . . .

The little yellow rubber elephant I always took with me
disappeared one day. First it was right on my bed, then it
was gone. That was one of the few times I cried.

Once my roommate was a young woman who had can-
cer. She had two little boys, three and five, and she talked
about them all the time. She was in the hospital for co-
balt treatment. When she was feeling low, she'd say, "I'm
going to die. I know it."

"I don't want to hear that kind of talk," I told her. "I
want you to get that idea out of your head."

She was about twenty-eight and I was about seventeen
then, and we got to be good friends. She was one of the
few real friends I ever made in the hospital. We wrote
each other every year at Christmas and sometimes in between. One Christmas I didn't hear from her, so I wrote
her sister. Her sister wrote back that she was dead. That
made me feel really terrible—it seemed unfair. She was
such a young woman and she wanted so much to live. I
hope her sister is taking good care of her boys.

Another time when I was in the hospital in Charlottesville, up and around and nearly ready to go home, Mama
had gone off to do some shopping. I began talking to a
very nice woman whose husband was in the hospital. She
was tired of just sitting there and she said to me, "Debbie,

would you like to drive out to Monticello and see Thomas Jefferson's home?"

Did I ever! It had been a real problem to me that this famous place was so close by, in the mountains just beyond Charlottesville, and I'd never gotten to see it. I'd already seen the beautiful white columns on the round library building which President Jefferson had built on the university campus. The hospital building, which also has those beautiful white columns, is not very far from the library.

Well, this nice woman called a taxi and we drove to Monticello. It was a wonderful, exciting place. The way it was built up on a hilltop reminded me of my church— nothing between the beautiful building and the sky. I liked President Jefferson's bed, which disappeared into the ceiling, and the dumbwaiters, which carried things upstairs and down, and all the gadgets and handsome furniture. It was the first time I'd ever been in a historic building. The sense of walking where famous people had walked gave me a strange feeling, as if I were part of history too.

Another time when I was in the hospital, there was a little girl there, a patient of Dr. Edgerton's, Beth, one of the many children who've come under the wing of the Debbie Fox Foundation. Beth's story is hard to believe. She was born with ocular hypertelorism and some other physical defects. But her main problem, the doctors told her parents, was that she was hopelessly retarded. The X-rays they took soon after birth showed that she had no front part of her brain. So her parents put her in a home for the retarded. Everybody thought she would just be a little vegetable, but she grew up and talked and played, and a kind Methodist minister took her into his kindergarten.

When she was about eight, her parents realized she wasn't retarded and brought her home. They heard about

Dr. Edgerton through one of the articles about me and took her to see him, and he did one of those big operations to bring her eyes closer together. When he opened her skull he found that her brain was misplaced down behind her nose, somewhat the way mine was, but much worse. He anchored her brain where it belonged, moved her eyes, and worked on her nose.

I met Beth in June 1976, when she was at the University of Virginia Hospital the second time for more work on her nose. Also at the hospital at that time was a boy named Charlie, who had been referred to Dr. Edgerton through the Debbie Fox Foundation. Charlie had one of those congenital deformities that required a great deal of corrective surgery on his face and mouth. Beth and Charlie are a lot younger than I am, and of course they'd had less experience with hospitals than I'd had. They were pretty worried, so I just sort of took them in hand.

I was in for what was, for me, minor surgery—grafting some hair from my scalp into a gap on my right eyebrow and shaping work on the tip of my nose and on my left upper eyelid. I had to stay in bed for only a short time. As soon as I was up, I walked around with a deck of cards in the pocket of my robe, and if I saw Beth and Charlie looking weepy, I played crazy ace or gin rummy with them. Some of those games got quite exciting. We'd shout and laugh a lot. I remember once a nurse came in and said, "Oh, the card sharks are at it again. May I interrupt to take some temperatures?"

I held up my hand to stop her. "Not till we finish this hand. The score is tied and this is it!"

The nurse waited. Charlie went gin. He couldn't talk because of his surgery, but Beth and I slapped him on the back, and he was one happy little boy.

The last time I heard about Beth, she was doing well

at public school, taking dancing lessons and playing the piano. I liked her a lot when I got to know her in the hospital, and since she lives in Tennessee, not too far from me, I hope we'll get together and talk about the miracle of surgery that has given us and so many others a real chance at life.

I learned recently that there are now cranio-facial clinics at half a dozen hospitals in the United States. Cranio-facial surgery has become a sub-specialty among plastic and reconstructive surgeons. There are medical meetings, exchanges of papers, new breakthroughs all the time. Operations that were considered impossible a decade ago are now performed almost routinely. Children with gross deformities of the skull and face are no longer thought hopeless. They can be made to look normal, or close to normal, just as I was. Each success seems to lead to another success. Dr. Edgerton was telling me recently that he hopes now to reach children with cranio-facial deformities early and correct their heads and faces when they are still very young.

We were sitting in his office in Charlottesville, a beautiful magnolia tree in full bloom outside his window. I was astonished to hear that he could operate on such young children. I remembered that Dr. Barnwell's plan for fixing my face was spaced out over my growing years. "Don't you have to wait until the children grow up?" I asked.

"We used to think that," Dr. Edgerton explained. "But we don't believe it anymore. The major work can usually be done in infancy or in the child's first three or four years."

"Then the child doesn't have to grow up deformed," I said in wonder.

"That's right, Debbie. If a child needs cranio-facial surgery, his parents should get him to a surgeon who does that kind of work as soon as possible. Then the boy or girl can have a more normal childhood and not have to experience the emotional problems of growing up with a severe deformity."

My mind flashed back to a little girl looking into a hand mirror for the first time and hearing the scream that tore out of her throat at the sight of her monstrous face. I remembered Mama's hand on my shoulder holding me back from running to join the children down by the creek. I could still feel inside me the intense longing to be with those happy, playing children.

Dr. Edgerton must have been reading my mind. "Debbie, do you have any idea how much you've done for other children?" he asked.

"I didn't do it," I told him. "You're the one who did it."

"We're all doing what we can," he said. "And it's not just for the children with the very severe deformities. We can help others, too. There are a lot of children who had operations for cleft palate some years ago. Those operations helped, but the children have grown up with sunken, caved-in faces. Now we know how to correct those faces by moving the bony structure in the central part of the face forward. Those children can now have a completely normal profile."

Dr. Edgerton went on to tell me that the kind of surgery he first did on me has led to new knowledge about the way the muscles around the eyes work and about the nature of bone growth and regeneration in the skull.

I listened to all this in wonder, but I had an important question to ask. I was remembering all the sad letters I'd gotten after my big operation from parents who were trying to find help for their afflicted children. "If parents have a child like me or like Beth or Charlie, how do they know where to turn?" I asked.

"Word is spreading among doctors because cases like yours get written up in the medical journals. The Debbie Fox Foundation has referred many hundreds of children to surgeons. My team has operated on more than a hundred and fifty children since you came to us. But you know, Debbie, I think there's something you can do. Have you thought about writing a book about what's happened to you? You could reach many thousands of people that way."

Dr. Edgerton saw I was smiling and asked why.

"I guess Mrs. Apple thought of the book first. Back when I was just a little kid she told me to start remembering things so I could write about them when I was bigger."

"There, you see, we both think it's a good idea."

That's how this book got started. I got in touch with the professional writer who'd written the magazine article about me, and she's worked with me, digging out the facts, talking to the doctors, and helping me put my story together. The book has kept me busy during this period when my time is still too broken into bits and pieces by my surgery for me to get and keep a regular job.

I had one of my most painful operations in the fall of 1977. Dr. Edgerton took some fatty tissue from the upper part of my right thigh and inserted it high on my cheeks to give my face better proportions and balance. My forehead had always seemed a little too wide and high. It turned out it really wasn't that at all. The lower part of my face had been too small and sunken in relation to my forehead. Once the cheeks were widened out, all the contours looked better.

But oh! what I went through before I could enjoy the results. The main trouble with that operation is that it hurt at both ends of me. For days I couldn't sit—except in extreme misery. Mama piled cushions on the chair under me, but nothing helped. I shifted position about every six seconds—there was no way to get comfortable until that place on my upper thigh healed. My face was more swollen than at any time since the big operation. As I squirmed and wiggled to get my bottom halfway eased, the nurses piled ice packs on my hot, swollen face.

Do you want to know how they put the ice on my face? I wish I had a picture of it! They filled a pair of transparent plastic gloves with crushed ice and propped the gloves on my face, and I peeked out as best I could between the icy fingers. I was in the hospital over Halloween that time. If I'd wanted to dress up for trick-or-treat, I wouldn't have had to do a thing. Those gloves with ice in them were better than any mask I've ever seen.

Once the swelling went down, I looked really good. When Mrs. Apple saw me, she said, "Well, Miss Debbie, do I see Claudette Colbert in front of me?"

I giggled at that, but I was delighted with the compliment. I've seen some of the old Claudette Colbert movies late at night on TV, and if anybody wants to think I now have cheeks as pretty as Claudette Colbert's—well, that's fine with me.


"Debbie, I really think it's time we did something about your hand. I'd like to go ahead. But you have to make the decision." It was Dr. Edgerton on the telephone.

"Let me have a week to think it over," I said.

"That's fine. You should think about it. If you have any questions, call and we'll try to answer them."

Mama got home from work very late that day. It was getting dark in the living room. I was curled up in the corner, in the big reclining chair next to the fireplace. I hadn't turned on the lights.

"What are you doing here in the dark?" Mama asked. She switched on the light. Then she saw I'd been crying. "What's the matter, Debbie? What's happened?"

Poor Mama! I guess she's always ready for another disaster. You can't blame her after what we've been through.

"It's my feet," I said. She looked at my feet. I was wearing my new red, open-toed sandals and stretched my feet out in front of me on the recliner.

"They look all right to me," Mama said.

"That's just it," I said, snuffling back the tears. "The hospital called. They want to do that toes-into-fingers thing. They want to do it soon."

Mama let out a big sigh. "You'll have to think about it, Debbie. You'll have to think about it real hard."

Mama smoothed back my hair with her hand as she walked by me into the kitchen. The touch of her hand on my hair reminded me of what I'd gone through when they cut off my hair that time at Johns Hopkins. This was the same all over again, only much worse. Hair grows back; toes don't.

I looked at my feet in these cute little red sandals. If some of my toes were cut away, my feet would be misshapen, deformed. How could I let that happen, even if the reward was a hand—of sorts?

Mama and I went through two weeks of misery and indecision. One day I'd be all for going ahead. It would be great to have a hand. I could drive better, cook better, stand a better chance of getting a job in an office.

The next day I'd say to myself, Debbie, you must be crazy. You'll have two messed-up feet and a not-very-good hand. Why go to the trouble?

"Do you suppose it'll hurt my balance if I let them take

my toes?" I asked Mama, not once, but about twenty times.

She finally said, "Didn't the doctor tell you to call if you had any questions? Why don't you call up and find out just what the operation means?"

I phoned the hospital and had a nice talk with one of the residents in plastic surgery. After that, I felt a lot better. He explained to me that digital transplants—that's what he called them—have been made possible by recent advances in micro vascular surgery. What that means is that the surgery is done under a microscope that enlarges the area by twenty times. Bones look as big as logs to the doctors; blood vessels look like ropes, nerves like wire coat hangers.

Using micro surgery, the surgeon can sew together veins, arteries, tendons, and nerves in such a way as to preserve function in the transplanted parts. Each toe to be moved would be taken out from where its bones began in the ball of the foot. There would be very little scarring on the foot. The main difference would be that my foot would look a little narrower in the toe area.

"What about my balance?" I asked. I don't know why I was so worried about balance. Maybe I'd heard Mrs. Huckaby talk about her problem of not being able to balance herself when she was a little girl, before she'd had her feet operated on.

"No problem about balance," the young doctor assured me. "When you stand or walk, you'd never know the difference."

That put my mind at rest a little. "But will the toes really work as fingers?"

"You should be able to move them and grasp things. They should be very useful to you."

"But won't they be too small and look funny?"

"No, they'll be about two and a half inches long—about as long as normal fingers."

"But how can that be? My toes are much smaller than that."

"Remember, I told you we'd take them from where the toe begins back in the foot. You've seen a skeleton, haven't you, or a skeleton costume? Well then, you know how long the toe bones look. That's what you'll get for fingers."

"And they'll work? I'll be able to wriggle them and move them?"

"That's the idea. That's what we're doing this for."

I thanked him and went back to my agonizing. There really wasn't a question anymore. I'd come this far. I'd been through more than half a hundred operations. Starting with almost nothing, I'd acquired a face that I no longer had to hide. It wasn't perfect; it never would be perfect. But it was a useful, presentable face. A stranger seeing me might think I'd been in some kind of automobile accident. A stranger would never imagine the distance I'd come.

I held up my funny little hand. If I had the fingers added now, a stranger might think it had been twisted in that same accident. A stranger would never dream I'd been born without a hand.

No, I couldn't stop now. I had to keep going forward. The tears were streaming down my face when the hospital called, but I managed a calm voice when I said, "It's okay. Set the date for the operation."

The operation took place in May 1978. It was far more dramatic and strenuous than I'd expected. I was in the operating room for more than twelve hours, and afterward

Dr. Edgerton told me that for the second time in my life I'd had surgery of a kind never performed before. Instead of taking a couple of insignificant little toes, the doctors actually removed the big toe of my right foot and attached it to my right hand. Nerve was tied to nerve, muscle to muscle, tendon to tendon, and, finally, the skin sewn together with the finest of stitches.

One of my big toes had become my missing thumb. But would it work? Mama and I went through days of painful waiting. The trees and shrubs outside my hospital room blossomed pink and white. I could hear the birds singing. But all I could do was look at my bandaged hand and foot and wonder what the world held for me.

I can tell you, I was scared, particularly when I first tried to stand up and take a step. I couldn't stand. I couldn't walk. I could feel a terror inside of me. "Mama," I cried, "I can't walk. I can't use my foot."

Mama's face was white with fear. Had we made the wrong decision? Had my worst fears come to pass?

Dr. Edgerton, calm and purposeful, as always, placed a soothing hand on my shoulder. "I think it's going to be all right, Debbie. I think it's going to work."

What Dr. Edgerton was pleased about was that my new thumb was a healthy pink color, meaning that the blood circulation was working. And I already had some feeling in it and could move it slightly. That was the miracle—the transformation of a toe into a working thumb. As for the sore, misbehaving foot—well, time would take care of that.

Although the final verdict is not yet in, I feel Dr. Edgerton was right once again. My new thumb still takes getting used to, but I'm pretty sure it's going to be a real working thumb. With exercise and some rehab training, my foot is getting back into shape. I'll be able to walk all right, but I'm not so sure about those pretty red sandals. I may not be able to wear them again, and that's a hard thing to accept.

Daddy and Mama told me I'd be fixed up and they kept their promise to me. God kept His promise. My face has been fixed and my hand soon will be. And in being fixed, I truly believe I became God's instrument for helping other children. That must have been His purpose all along.

Just the other day Mrs. Apple and I were talking about all we've been through together, and she said, "Debbie, I can't help thinking that there *was* a purpose in everything. There was a purpose in your being born the way you were. There must have been a hidden purpose in Dr. Barnwell's death, in what happened to the Atlanta doctor, in your getting to Dr. Edgerton, and in his finding out through you that all those hopeless children can be given new faces and normal lives.

"You know, Debbie, you've been a blessing to everyone whose life you've touched."

I blushed when she said that. I know she's exaggerating. It's not I who have blessed others. It's God who's fulfilling His purpose, for as we sing in the hymn, "Nothing is too difficult for God to do."

What am I going to do with the life I have been given? I haven't really made up my mind yet. I've been concentrating so hard on getting my face and my hand fixed that I haven't come to any final decisions. My church has meant so much to me that I may go into religious work. Or I may take training that will let me work in a hospital. I think I'd like to work in a hospital with children and with the handicapped. I know what it means for a small child to be in a strange hospital room, separated from parents and frightened about what's going to happen. I could do a lot to help those children. At the same time, I could pay back the doctors, the nurses, the technicians, and everybody else for giving me the face that I was denied when I was born.

I want to do whatever I can to help the Debbie Fox Foundation for the Treatment of Cranio-facial Deformities in its work. Since it started, the Foundation has received about fifty thousand dollars in donations, referred hundreds of children to doctors and medical centers, encouraged research, and helped scores of families pay the heavy travel and living expenses connected with corrective surgery. By the way, contributions to the Foundation can be sent to P.O. Box 11082, Chattanooga, Tennessee 37401.

Some day I'd like to be married and have children of my own. I often dream of that. I've been told it's perfectly safe because my condition is not anything that could be inherited. What happened to me was a cruel accident that God shaped to His purpose; there's no reason why I shouldn't have entirely normal children. I would like to have a husband who loves me. I would like to have a happy marriage and children to love and care for. That's what I dream about.

If I have a career, I want it to be visible. I want to show my face. Showing it will honor the doctors who gave me my face. It will honor my parents, who believed, even in the darkest days, that I would be made whole. It will please me because I've had too much of hiding in my life. Most important, it will give meaning to my long search for a face. It will tell those who look at me that the handicapped and the deformed must not be kept locked away from view. *We must come out*, into the mainstream of daily life.

And those who look at us must accept us as full partners. Some of us may be hard to look at. Some of us may have trouble getting around. Some of us may not see or hear. We may have difficulty expressing ourselves. Our bodies may be twisted. But we have much to offer. We have overcome disabilities that have crippled not only our bodies but our spirits, too. We have moved beyond our physical deformities. Through surgery, science, and faith, we have come into the light. And we must give witness to God's love for us.

Whatever path I choose, I know that God would want me to serve and give to others, because, as I wrote years ago in one of my poems,

> This has been done for me
> And I have been set free.